ILLUSTRATED

SCIENTIFIC AND DESCRIPTIVE

CATALOGUE

OF

ACHROMATIC MICROSCOPES,

MANUFACTURED BY

J. & W. GRUNOW & CO.,

NEW HAVEN, CONN.

PRICE 30 CENTS.

NEW HAVEN:

T J. STAFFORD, PRINTER.

1857.

PREFACE.

----◆◆◆----

In presenting to the public this Illustrated and Descriptive Catalogue, it is our intention to furnish, especially to those living at a distance, the means of becoming acquainted with the various forms and qualities of the microscopes and microscopical apparatus which we manufacture.

Having been engaged almost exclusively, for a series of years, in the manufacture of this important instrument, we have devoted great care and attention to improving the structure of the mechanical parts, as well as to perfecting the quality and efficiency of our object-glasses, and the optical parts in general.

The extensive patronage with which we have been favored affords sufficient evidence that in our efforts to deserve the confidence of the scientific public we have not been without success.

But while it has been our object to bring to perfection the branch of practical optics to which we have devoted ourselves, we have been constantly adding to our facilities for manufacturing the various apparatus connected with the microscope. By the introduction of machinery especially adapted to our purposes, and by a judicious division of labor, we have endeavored to secure the most finished workmanship in every part, at such prices as shall place the most perfect instruments within the reach of all the lovers of science and students of nature.

We are daily receiving inquiries from persons who desire information and advice in regard to the selection of microscopes adapted to their particular purposes. It frequently happens that persons making these inquiries have had but little opportunity to become acquainted with the theory and uses of microscopes made in the most recent and improved style. Properly to reply to all these inquiries would absorb too much of our time, which we desire to devote exclusively to the more delicate and important labor of our art. We have therefore decided to publish in our catalogue such information as shall enable every one to understand the philosophy and structure of the most improved microscopes, and to judge of the qualities of different microscopes that may be offered to their patronage.

So rapidly has the microscope been improved within a very recent period, that instruments which were made but a few years since have become of little value, compared with the improved achromatic miscroscopes which are now made by the best opticians of England and America.

This rapid advance of improvement has caused a large stock of microscopes of inferior quality to be thrown into the market at very low prices, and those who are not aware of the great superiority of the achromatic microscope, as at present constructed, are induced to purchase cheap, but *inferior*, instruments. With this species of trade it is not our purpose to compete. But to those who desire to obtain microscopes of the best construction, combining the most recent scientific improvements and fitted for the prosecution of the highest order of scientific inquiries, we offer our microscopes with confidence that wherever their merits are known they will give ample satisfaction to purchasers.

J. & W. GRUNOW & Co.

New Haven, Conn., Oct. 1, 1857.

CONTENTS.

ACHROMATIC MICROSCOPES.

CHAPTER I.

THEORY OF THE MICROSCOPE

SECTION		PAGE
1.	Introduction,...	1
2.	Simple Lenses, . ..	1
3.	Foci of Lenses,.	1
4	**Aberration,** ...	2
5	Spherical Aberration, ...	2
6	Amount of Spherical Aberration,	3
7.	Proportionate Curvature,	4
8	Negative Aberration, ...	4
9.	Aberration of Sphericity Curvature of Image. .	4
10	Chromatic Aberration,...	5
11	First method of diminishing Chromatic Aberration,	6
12.	Second " " " " "	6
13	Achromatism, ...	6
14.	Combined Lenses, ...	7
15	Angular Aperture, ...	8
16.	Oblique Illumination,	9
17	Lister's Discoveries,........	9
18	Essential requisites of good Object-Glasses,.	9
19	Mr Lister's two Preliminary Propositions,.	10
20.	Principles discovered by Mr. Lister, ...	10
21	Aplanatic Object-Glasses, ...	11
22	Superiority of English and American Object-Glasses,	12
23	Compound Achromatic Microscope, ...	12
24	Positive Eye-piece,.	12
25	Negative Eye-piece	12
26	General view of Compound Achromatic Microscopes, ...	13
27	Enlarged section of Achromatic Microscope,.....	15
28	Action of Negative Eye-piece,	16
29	Advantage of Over-correcting Object-Glass, ...	17
30	Negative Eye-piece nearly Achromatic, .	17
31	Use of the term Achromatic Objective,	18
32	Aberration produced by Glass Cover,.....	19
33	Objective Corrected for Glass Cover,	20
34	Eye-pieces of different Powers, ...	21

SECTION		PAGE
35	Draw-tube,.	21
36	Names applied to Object-Glasses,.	21
37	Foci of Higher Powers inconveniently near to Objective, ... 22

CHAPTER II.

MECHANICAL PORTION OF THE MICROSCOPE.

38.	Modern Improvements, ... , ...	23
39.	Base,.............	23
40,	Arrangement for Inclining the Instrument, ...	23
41	Stage,.............................	24
42	Adjustment of Focus,.....	24
43	Simplicity and Facility of Adjustment,..	24
44	Description of Microscope Stands, ...	24
45	No 1, Educational Microscope, ...	25
46	No 2, Student's Microscope, ... 27
47	No 3, Student's Microscope, ... 28
48	No 4, Student's Larger Microscope,. 29
49	Stage movable by Rack and Screw,... 30
50	No 5, Another form of Student's Microscope, ... 31
51	Chevalier's Prismatic Body, ... 32
52	Prof Bailey's Indicator Stage, ... 33
53	No 6, Portable Microscope, . . . 35
54.	No 7, Large Microscope, ... 37
55	Chemical or Inverted Microscopes. 38
56	No 8, Simple form of Inverted Microscope, ... 39
57	More complete Inverted Microscope, ... 40
58	**Object-Glasses,**. 40
59	Rules for Adjusting Object-Glasses, ... 42
60	Delicacy of Adjustment for Thin Covers, ... 42
61	Second Class Objectives, ... 43

CHAPTER III

ACCESSORY APPARATUS.

62	**Micrometers,** 45
63.	Glass Stage Micrometers Mounted in Brass,....... 15
64	The Cobweb Micrometer, ... 45
65.	Ross' Eye-piece Micrometer, ... 47
66	Jackson's Micrometer,. 47
67.	Comparative Merits of Micrometers,.. 48

SECTION   PAGE

68   Dr. White's Micrometer, ... 49
69. Prof J Lawrence Smith's Goniometer and Micrometer, ... 50
70   Method of using the Goniometer, ... 51
71   Value of Lines in Eye-piece Micrometers, ... 51
72   Method of using the Micrometer, ... 53
73   Fraunhofer's Stage-Screw Micrometer, ... 54

Camera Lucida.

74   Wollaston's Camera Lucida, ... 56
75   Nachet's Camera Lucida, ... 56
76   Soemmering's Steel Speculum, ... 57
77   Using the Camera, ... 57
78   Camera Lucida applied to Micrometry, ... 58

Miscellaneous Apparatus.

79   Movable Diaphragm Plate, ... 58
80   Bull's-Eye Condenser, ... 59
81. Smaller Bull's-Eye Condenser, ... 60
82   Achromatic Condenser, ... 61
83   Nachet's Prism for Oblique Illumination, . 62
84   Lieberkuhn Speculum, ... 63
85   Erector, ... 64
86. Orthoscopic Eye-piece, ... 64
87   Compressor, ... 65
88   Animalcule Cage with Screw, ... 65
89   Simple Animalcule Cage, ... 66
90   Stage Forceps Hand Forceps, ... 66
91   Frog Plate, ... 66
92   Machine for cutting circles of thin glass, ... 67
93   Instrument for making cells of gold size, ... 69

CHAPTER IV.

POLARIZED LIGHT AND ITS APPLICATION TO THE MICROSCOPE.

94. Theories of Light, ... 70
95. Double Refraction and Polarized Light, ... 70
96   Polarization by Reflection, ... 71
97   Polarization by Refraction, ... 72
98. Polarization by bundles of thin plates, ... 71
99   Double Refraction, ... 74
100. Double Refraction of Iceland Spar, ... 75

SECTION   PAGE
101   Polarization produced by Double Refraction, ... 77
102   Nicol's Single Image Prism, ... 77
103   Common and Polarized Light Contrasted, ... 79

Theory of Polarized Light.

104   Undulatory Theory,.... 79
105. Illustrations of wave motion, ... 80
106. Polarization illustrated by resultant motion, ... 81
107   Polarizing effect of Iceland Spar,....... 82
108. Familiar Illustrations,........... 83
109   Partial Polarization,........... 83

Polarized Light applied to the Microscope.

110   Polarizing Apparatus, ... 85
111   Tourmaline Plates, ... 86
112   Herapathite, ... 87
113. Value of Polarized Light in Microscopic Investigations, ... 87
114   Colored Polarization, ... 87
115   Cause of Colors produced by plates of selenite, ... 88
116   Method of Varying Colors,.... 88
117   Selenite Stage, ... 90
118   Polarizer with Revolving Selentine Carrier,... 90
119   Delicate structures viewed by colored polarized light, ... 91
120   List of Objects for Polariscope, ... 92

CHAPTER V.

PRACTICAL DIRECTIONS.

121   Care of the Microscope, ... 94
122. Illumination, ... 95
123   Choice of a Microscope, ... 95
124   Qualities of Object-Glasses,...   96
      1   Defining power,...   96
      2   Resolving power, ... 97
          Method of measuring angular aperture, ... 97
      3   Flatness of field,... 97
      4   Depth or Extent of Definition, ... 97
      PRICE LIST, ... 99

ACHROMATIC MICROSCOPES.

CHAPTER I.

THEORY OF THE MICROSCOPE.

1. **Introduction.** In calling public attention to the claims of Compound Achromatic Microscopes, and their great superiority over all others, it is necessary to explain the defects of simple lenses, the general structure and action of achromatic lenses, and also the entire optical and mechanical arrangement of the most improved *Compound Achromatic Microscopes*. It is presumed that every reader of this catalogue understands, or can easily learn from ordinary works, the simple elements of optics, and the structure of simple and compound miscroscopes formed of simple lenses

2. **Simple Lenses.** The names applied to the various forms of simple lenses sufficiently explain their structure, and are generally understood. All lenses are supposed to be bounded by plane, or spherical, surfaces ; for although some other forms would be desirable if they could be made with sufficient accuracy, it is found in practice that only plane and spherical surfaces can be wrought so perfectly as to render them available for optical purposes.

3. **Foci of Lenses.** A *plano convex lens* E E, Fig. 1, (or any spherical lens thicker in the centre than at its edges,) refracts parallel rays of light L L, to a point F, called its

1

principal focus. If a pencil of light diverging from a point behind the lens, as F', more distant from the lens than the

Fig. 1.

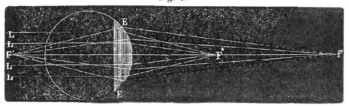

principal focus, falls upon the lens, it will be refracted to a focus f, also more distant than the principal focus. When F' approaches the lens, f will recede from it, and *vice versa*, hence these two are called the conjugate foci of the lens. The relative distances of the conjugate foci have a certain relation to the distance of the principal focus F.

4. **Aberration.** In a perfect lens the above statements would be strictly correct, but in all spherical lenses it is found, on careful examination, that the rays falling upon the lens at different distances from the centre, do not all meet in a single point, but are subject to two different causes of error, called *spherical* and *chromatic* aberration, which will now be explained.

Fig. 2.

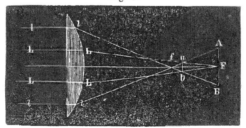

5. **Spherical Aberration.** Let F, Fig. 2, represent the focus of two parallel rays L L, near the centre of the lens; then if *l l* represent the rays falling upon the borders of the lens, they will be so refracted as to meet at a point f, con-

siderably nearer to the lens than the point F, where the central parts of the pencil converge. The distance F f, is called the *longitudinal spherical* aberration, and *a b*, the smallest diameter of the converging pencil, is called the *lateral spherical aberration.*

6. It is easily seen that if the plano convex lens is turned with its convex surface towards parallel rays, the points where the lateral rays undergo refraction would then be in advance of the positions where the rays, passing near the centre, are refracted, and consequently the spherical aberration would be diminished by the difference between the thickness of the lens at the centre and the thickness at the border. The difference is really much greater than this, for if the plane side of the lens is turned towards parallel rays, as shown in Fig. 2, the longitudinal spherical aberration is about $4\frac{1}{2}$ times the thickness of the lens, but if the opposite face is towards parallel rays, the aberration is only about $1\frac{17}{100}$ times its thickness.*

A double convex lens has the least spherical aberration when the two faces have their radii of curvature in the proportions of about one to six. Such a lens with its flatter surface

---

* The spherical aberration of a lens varies with the refractive power of the material of which the lens is made

The proportions here given are taken from Lardner's Optics Other writers state them somewhat differently

According to Prechtl, if n=index of refraction, r=the radius of curvature of the anterior surface of the lens, and R=radius of posterior surface, then for parallel rays the form of least aberration will be expressed by the equation

$$\frac{r}{R}=\frac{4+n-2n^2}{n(2n+1)}$$

Then if the index of refraction equals $1\frac{1}{2}$, the form of least aberration will be obtained if the two surfaces have their radii as 1 to 6, the side of deeper curvature being towards the parallel rays. If the spherical aberration of such a lens in its best position be reckoned as unity, the aberrations of other lenses will be as follows

Plano convex with plane surface towards distant objects, 4 2.
     "   "   " convex surface towards distant objects, 1 081

Plano concave the same as plano convex

Double convex or double concave with both faces of the same curvature, the aberration will be 1 567.

turned towards parallel rays, has its spherical aberration about $3\frac{1}{2}$ times its thickness, while in the reverse position the aberration is only about $1\frac{7}{100}$ times the thickness of the lens.

7. Proportionate Curvature. It is thus seen that the spherical aberration of a lens may be considerably reduced, by giving a proper proportion to the respective curvatures of its two surfaces, and by turning the more convex surface towards parallel rays, or the rays that are nearest parallel.

The thickness of a lens being very nearly as the square of its diameter, a lens of small diameter will have only about one fourth as much spherical aberration as a lens of the same curvature with double its diameter.

Hence, if only the central portion of the lens is used, the aberration will be still further diminished. This plan is adopted in all the common and cheap microscopes, but the amount of light transmitted is very small, and other imperfections result, which will be understood when we treat of *angular aperture*. (See 15, 16, 17.)

If the refraction is performed by two lenses of shallow curvature, the aberration is less than if the same amount of refraction takes place in a single lens of deep curvature.

8. Negative Aberration. A concave lens has the same amount of spherical aberration as a convex lens, but it takes place in an opposite direction, and is therefore called *negative aberration*.

9. Aberration of Sphericity: Curvature of the Image. When a flat object is viewed through a single lens, so placed that the central portion is clearly seen, the borders of the object appear indistinct, and the lens must be brought still nearer in order to view the lateral portions clearly. This effect takes place chiefly because the lateral portions of the object are more distant from the *optical centre* of the lens than the central portions, and partly because the refractive power of the lens is exerted more strongly on pencils of light, passing obliquely through it, than on those passing so that their axes coincide with the axis of the lens. For the same reason an image formed by a lens appears curved towards the lens by which it is formed, and when this curved image is viewed by another lens, as in the *common* compound microscope, the distortion of the image is still further increased.

This effect, called aberration of sphericity, or curvature of the image, is not diminished by contracting the aperture of the lens, nor even by dividing the refraction between two convex lenses, so placed as to act together as a single lens; but when the longitudinal spherical aberration is corrected, the aberration of sphericity, by proper arrangements, may be corrected also, so that every part of a flat object shall be distinctly defined upon a flat field. This quality, called *flatness of field* is very essential to a good microscope.

10. **Chromatic Aberration** is another error which arises in the use of a single lens. Whatever be the form of the lens, the material of which it is composed does not act uniformly upon the differently colored rays of which white light is composed, but separates each ray of white light, falling obliquely upon its surface, into the colors of the prismatic spectrum.

Fig. 3.

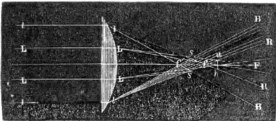

Let L L, *l l*, Fig. 3, represent rays of white light falling upon a plano convex lens. The ray *l l*, being nearer the border of the lens is strongly refracted, and the blue ray *l* B diverges widely from the red ray *l* R, and generally, the blue rays are brought to a focus nearer to the lens than the red rays.

The rays L L, falling very near the axis of the lens, are almost perpendicular to the refracting surface, and hence the colored rays of which the white light L L is composed, are but slightly separated from each other. From this we see that the chromatic dispersion is small near the centre of the lens, and like spherical aberration, increases very rapidly towards the borders of the lens.

In this figure, as in Fig. 2, F f represents the longitudinal spherical aberration, while the chromatic aberration extends over the entire space from F to f'. The chromatic aberration is therefore in the same direction as the spherical aberration.

Here the narrowest part of the pencil, and hence the most available focus, is not at *a b*, but at *s s*. In this case, as with spherical aberration, if the lens were placed with its convex surface toward the parallel rays, it is obvious that the longitudinal chromatic aberration F f' would be diminished, but the actual dispersion of the colors in each ray of light would remain unchanged, except as the change of position causes a slight alteration of the refractive power of the lens.

11. **The first efforts to diminish the Chromatic Aberration** of a single lens consisted in reducing its diameter, as is seen in the cheap microscopes so commonly found in the market: in these the orifice through which the light passes is exceedingly small, and consequently the object appears but feebly illuminated. In the compound microscope of such construction only a low magnifying power can be used.

12. **The second method of diminishing the Chromatic Aberration** consists in employing two or three lenses of shallow curvature placed close together, by which means the chromatic and spherical aberrations are made as small as is possible with lenses composed of a single kind of glass. This form of lenses, called doublets and triplets, is also seen in cheap compound microscopes of French and German manufacture. But in this form a considerable amount of chromatic aberration still remains, even when the diameter of the lenses is quite small.

13. **Achromatism.** From the preceding considerations it is evident, that it is of primary importance, in the construction of a really efficient microscope, that the chromatic and spherical aberration should be *entirely corrected*, but that no such correction can be effected in a single lens.

Near the middle of the last century, John Dollond of London, by combining a double convex lens of crown glass with a concave lens of flint glass, succeeded in constructing an object-glass for the telescope in which the greater dispersive power of the flint glass served to correct the chromatic aberration of the convex lens of crown glass, while a considerable portion of the refractive power of the crown glass remained as the efficient power of the compound lens. This principle was soon thoroughly established as in every way successful for forming achromatic lenses for the telescope. Efforts were made in the early part of the present century to improve the microscope in the same manner. In applying this principle to the microscope several difficulties were encountered. First, it was found extremely difficult to work the curves of such small lenses with sufficient accuracy to insure freedom from chromatic aberration. Secondly, while lenses for the telescope of large diameter constituted but very small segments of spheres, the lenses requiring long foci, and being adapted to receive rays very nearly parallel, lenses for the microscope, receiving light radiating from a point very near the lens, when made of comparatively moderate diameter constituted *very much larger* segments of spheres than the lenses of telescopes. Thirdly, it was found that when the aberrations of a convex lens of crown glass were corrected by a concave lens of flint glass, if the diameter of the lens was enlarged beyond very moderate limits, the correction for the borders of the lens became too great in proportion to the central portion, so that there seemed to be a limit, and that a very small one, to the available aperture of achromatic lenses constructed on this principle. So great, and apparently invincible, were these difficulties, that as late as 1824, such philosophers as Biot and Wollaston predicted that the compound achromatic microscope could never be brought to the same degree of perfection as the achromatic telescope.

14. **Combined Lenses.** To overcome these difficulties, different opticians combined two or more compound lenses, each made separately as nearly achromatic as possible. By this means a higher magnifying power was obtained, and a larger angular pencil of light was transmitted, though considerable

light was lost by reflection from the numerous refracting surfaces employed. The lenses of each achromatic combination were therefore next cemented together, greatly diminishing the loss of light by reflection. The compound lenses, used for the formation of doublets and triplets, being each separately made as nearly achromatic as possible, could be used singly for low powers, and two, three, or even four of nearly the same focus could be screwed together in a tube, constituting a compound objective of high magnifying power.

French and German achromatic objectives are still made on the principle of allowing each achromatic lens to be used separately, though the several combinations of which an objective is composed have different magnifying powers when used alone.

15. **Angular Aperture.** It is necessary to explain distinctly what is meant by *angular aperture*, referring not simply to the diameter of the lens, but to the *angular* divergence of the extreme rays of the pencil of light, which a lens is adapted to receive. It depends on having the diameter of the lens large in proportion to the distance between the lens and the object, so that a lens of short focus may have a very large *angular aperture*, though its absolute diameter is small.

The amount of light by which any point of an object appears illuminated, depends on the angular dimensions of the transmitted pencil, or, which is the same thing, the angular aperture of the object glass.

If the number of rays of light from any object be insufficient, it cannot be seen even though we employ a microscope for the purpose. With high magnifying powers the object appearing greatly enlarged, the light is spread out over a large surface, and, unless the amount of light is proportioned to the magnifying power, the object appears dark and imperfectly illuminated.

With a large angular aperture many delicate markings appear, which are quite invisible with a smaller aperture. A greater contrast is seen between different parts of the object with a large angular aperture, occasioned probably by the fact that the lens takes in a larger angular pencil of light from elevations, than *can issue* from minute depressions; differences in the density of different parts of an object produce a similar effect.

16. **Oblique Illumination.** When an object cannot be illuminated by a large angular pencil of light, nearly the same effect may be produced by employing very oblique illumination, provided the object-glass has a large angular aperture, so as to enable it to take in a very oblique pencil falling in a direction to be taken up by one side of the lens only. This may be illustrated by the well known fact, that the numerous crags and peaks of distant mountains are better distinguished by the oblique rays of the setting or rising sun, than by the more direct meridian rays ; and for the same reason astronomers can see and measure the mountains in the moon better when it is *new*, or at the *quarter*, than when it is *full*.

17. **Lister's Discoveries.** Among those who have contributed to the improvement of the compound achromatic microscope, first and foremost stands the name of Joseph Jackson Lister, Esq. This observer, in 1830, presented to the Royal Society a paper, in which he pointed out certain newly discovered properties of achromatic lenses, by taking advantage of which object-glasses could be constructed, consisting of three compound lenses, each having its aberrations more or less corrected, by whose combined action a very large angular pencil could be transmitted, and admitting at the same time, with ease and certainty, of entire correction over the whole field, perfect definition extending over an entirely flat field.

So valuable were the discoveries of Mr. Lister, that his paper has formed the basis of all the improvements in the compound achromatic microscope which have been made up to the present time. Stimulated by Mr. Lister's discoveries, practical opticians have carried the improvement of the microscope so far that theory itself seems to point to but little further improvement to be desired, and the compound achromatic microscope has reached a degree of perfection, hardly equaled by even the most recent improvements of the achromatic telescope.

18. **The essential requisites of a good Object-Glass,** for the compound microscope, as stated by Mr. Lister, are: First, the transmission of a large focal pencil, free from all aberration. Second : that the field of view should be flat and well defined

throughout.   Third : that the light admitted should, as much as
possible, be only such as goes to form the image, and it should
not be intercepted or diffused over the field by too many
reflections.

19. **Mr. Lister's two preliminary propositions are :**
First : that if a plano concave lens of flint glass is employed to
correct the aberrations of a double convex lens of crown glass,
the correct centering of the lenses is more easily effected, and
accurate workmanship for a short focus is much simplified ;
though other forms may be employed for special purposes.   Second : that the concave and convex lenses should be cemented together with some substance permanently homogeneous, and the
two curves should be identical in form, and pressed close
together, so as to leave between them but a very thin layer of
cement.   By cementing the lenses in this manner, the loss of
light by reflection is diminished by nearly one half.   Third,
that the compound plano convex lens thus formed, with curves
which render it nearly free from aberration, should always be
used with its flat surface towards the object to be examined.

20. **The principles discovered by Mr. Lister** are briefly
stated as follows :

In lenses formed as above mentioned, with a plano concave
lens of flint-glass cemented to a double convex lens of crown,
rendered achromatic by proper adjustment of the curves of the
two lenses, there is some point not far from the principal focus,
on the plane side of the compound lens, and situated *in its
axis*, from which light falling upon the lens is transmitted free
also from spherical aberration, and emerging either nearly
parallel, or tending to a conjugate focus within the tube of a
microscope.   If the radiant point is brought nearer to the lens,
the spherical aberration will be over-corrected ; but if the radiant continues to approach the lens, another point will be found
for which the spherical aberration is again exactly balanced.
For every radiant point still nearer to the lens, or more distant
than the first point, the spherical aberration will be under-
corrected.   The two radiant points for which the lens is perfectly corrected, both for chromatic and spherical aberration,

are called the *aplanatic foci*. When the longer aplanatic focus is used, the marginal rays of an oblique pencil (from a point on one side of the axis) are distorted so that the objects seen in the borders of the field appear fringed with a coma extending outwards, while the contrary effect of a coma directed towards the centre of the field is produced by the rays from the shorter focus The correction of chromatic aberration, like that of the spherical, tends to excess in the marginal rays.

21. Aplanatic Object-Glasses. These principles afford the means of destroying, with the utmost certainty, both aberrations in a large focal pencil, by combining two or more achromatic lenses in a single object-glass. The rays of light from an object are received by the anterior combination from its shorter aplanatic focus, and are transmitted to a second achromatic lens, of such form, and so placed, as to receive the rays in the direction of its longer aplanatic focus. If the lenses are fixed at this distance, the radiant point may be moved backward and forward, as required to increase or diminish the length of the microscope, without disturbing the balance of the corrections; since the motion of the radiant point produces opposite and equivalent errors in the two compound lenses. Slight errors in color may be destroyed in the same manner by their opposites, and thus we not only acquire fine correction for the central ray, but all coma of oblique pencils is destroyed, and the whole field is rendered beautifully flat and distinct.

In the application of Mr. Lister's principles, in order to enlarge the angular aperture as much as possible, it is found better (as Mr. Lister himself suggested) to retain in the anterior combination a certain amount of positive aberration, to be corrected in the posterior combinations, the proportionate dimensions of which are somewhat varied to secure the best effect in the entire compound achromatic objective. Sometimes, also, in objectives of high power, the anterior and posterior combinations are each made to consist of three lenses, while the middle combination has but two; but each combination is specially calculated for the place it is to occupy, and more or less corrected by itself, as is found best to secure the most delicate performance of the entire combination. Such a combination of lenses is called an *aplanatic object-glass* or *objective*.

22 **Cause of Superiority of English and American Objectives.** The best English and American opticians seek the highest perfection in each objective, and hence their glasses cannot be separated and arranged in new combinations. While opticians on the continent of Europe, almost without exception, still make their achromatic object-glasses on the old plan of separating the combinations, using the several glasses separately, or uniting them in new combinations.

Although French and German achromatic object-glasses may thus be furnished at a low price, their performance cannot equal that of the best English and American glasses, which will always be preferred by those who understand the philosophy of the microscope, and are capable of judging of its qualities.

23. Compound Achromatic Microscope. In using the achromatic object-glass for microscopial purposes, it is usually combined with another instrument known as the *eye-piece*, the two, together, constituting the compound—achromatic—microscope. The eye-pieces used for the microscope are the same as are employed for the astronomical telescope, and consist of two kinds.

24. The Positive Eye-piece consists of two plano convex lenses, with their convex sides turned towards each other, and set at such a distance that their compound focus is in front of the first lens. An eye-piece of this construction has less achromatic and spherical aberration than any single lens of the same power, but as it causes some renewed coloring and distortion of the image formed by the object-glass, it is only used in the microscope for special purposes.

25. The Negative Eye-piece, invented by Huyghens, for the telescope, is so named because it requires the image to be formed behind the first lens, the second, or eye-lens, only being employed to view the image. Two plano convex lenses are placed with their plane surfaces towards the eye, at a distance from each other, equal to half the sum of their focal lengths, and with a stop or diaphragm placed midway between the lenses. Huyghens intended this eye-piece to diminish the spherical aberration, like the positive eye-piece, and especially to enlarge the field of view, both valuable qualities, but he was not aware of the most important excellence of his invention. It was reserved for Boscovich to show that he had by this important arrangement accidentally corrected a great part of the chromatic aberration, as will be shown hereafter. (See 30.) The negative eye-piece is therefore the one generally employed for the microscope.

26. **A Section of a modern Compound Achromatic Microscope,** is shown at Fig. 4, where O is an object, and above it is seen the triple achromatic objective. The lenses E E, and F F, constitute the negative eye-piece invented by Huyghens. The plano convex lens E E, is called the eye-glass, F F is the field-glass, and between them, at B B, is a dark stop or diaphragm.

The course of light is shown by the three rays drawn from the centre, and three from each end of the object O; these rays, if not prevented by the lens F F, or the diaphragm at B B, would form an image at A A; but as they meet with the lens F F, in their passage, they are converged by it, and meet at B B, where the diaphragm is placed to intercept all

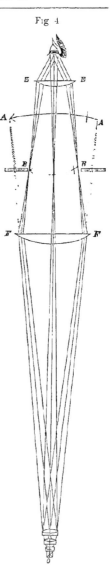

Fig 4

the light, except that required for the formation of a perfect image, and to limit the field of view to such an aperture as will be well defined when viewed with the eye-lens E E.

The image formed at B B, is further magnified by the eye-lens, as if it were an original object. The triple achromatic object-glass, constructed on the principles discovered by Mr. Lister, though capable of transmitting large angular pencils, and corrected as to its own errors of spherical and chromatic aberration, would, nevertheless, be incomplete without some special adaptation to prevent renewed aberrations distorting the image as transmitted and viewed by the eye-piece.

The property of the negative eye-piece, pointed out by Boscovich, most admirably meets this condition; for although it would not be free from aberrations when used alone, yet, when used as an eye-piece, in connection with an objective, its effect upon converging *pencils* is such that the aberrations produced by the field-lens are very nearly balanced by opposite aberrations in the eye-lens, resulting from the fact that it is situated on the opposite side of the image formed between the two lenses, and because the light from the object falls only upon those parts of the field-lens F F, which are best adapted to transmit it free from error.

27. Enlarged Section of the Compound Microscope. A more complete view of the action of the several parts of the compound achromatic microscope, is given in Fig. 5, the lenses being represented on an enlarged scale. A A, M M, P P, represent the three compound lenses of which the achromatic objective is composed. F F is the field-lens, and E E the eye-lens of the negative eye-piece.

Three rays drawn from the centre of the object O, and three from each extremity, show the course of both direct and oblique pencils. It is impossible in the space allowed to the figure, to show the separate action of each concave and convex lens, but only the action of the objective considered as a whole.

First: the axial rays, both of direct and oblique pencils, cross at some point which constitutes the *optical center* of the compound objective, and emerge from the posterior lens in the same direction they pursued on leaving the object; the lateral displacement which they undergo is too small to be shown in the figure, and is altogether unimportant.

Second : the extreme rays from each pencil cross each other in the borders of the objective.

Third: the rays issue from the posterior combination slightly over-corrected for color, so that if no other lens intervened a blue image would be formed at B B, and a red image at R R, while other images, with intermediate colors, would fill up all the space between B B and R R.

In a single convex lens, as we have seen in § 10, the blue image is formed nearer to the lens than the red.

This slight over-correction of the object-glass, really much less than shown in the

Note.—The course of rays shown is not strictly accurate, but such as to show correctly the principles referred to in the description.

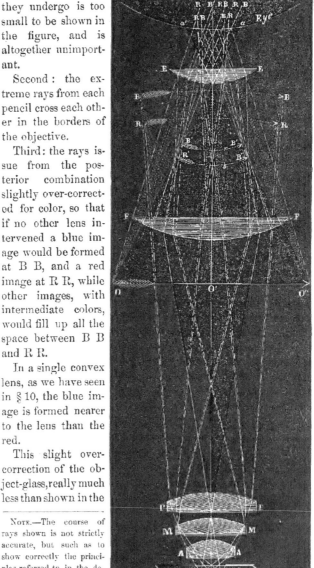

Fig. 5.

figure, is purposely produced, to balance the small amount of aberration which remains otherwise uncorrected in the negative eye-piece. The aberration of sphericity connected with the object-glass, is slightly under-corrected, as is shown by the images B B and R R, which are turned with their concave sides towards the object-glass.

28. **Action of the Negative Eye-Piece.** The field lens F F, bends the lateral pencils inwards, forming images nearer and smaller than would have been formed without its action. This action of the field-lens diminishes somewhat the magnifying power of the instrument, but it enlarges the field of view, and it is hence called the field-lens.

Secondly : the field-lens refracting the blue-rays more strongly than the red, the blue and red images are brought nearer together, as shown at B' B', R' R'.

Thirdly : the blue rays being bent out of their course by the field-lens more than the red, the blue image becomes so much smaller than the red that both images are seen in the same direction, as shown by the lines of sight, $a\ O$, and a' O'', Fig. 5, which, by their intersection at the optical centre of the eye-lens, form the visual angle under which the magnified image of the object is seen, as though situated at O O' O''.

Fourthly : the field-lens being thicker in its centre than at the edges, and the rays from the centre of the object all falling upon the central portion of the field-lens, while the pencils from the extremities of the object fall upon the borders of the field-lens, the curvature of the image is reversed, and the images B' B', R R', have their concave surfaces turned towards the eye-lens, and are in the exact condition to appear as a single straight image at O O' O'', when viewed through the eye-lens E E.

Fifthly : the lines $a\ O$, and a' O'', passing through the optical centre of the eye-lens, and through the extremities of the real images B' B', and R' R', if extended back to the limit of distinct vision, (which is generally reckoned at ten inches,) show the direction and apparent size of the magnified image, as it appears to the eye.

Sixthly: the red and blue rays, after being refracted by the eye-lens, arrange themselves in such relation to the lines $a\,O$ and $a'\,O''$, that they all appear to proceed from a single straight image, entirely free from both chromatic and spherical aberration. The rays $B\,O$ and $R\,O$ do not actually coincide in position, but they appear to emanate from the same point, while other rays, emanating from the same point in the object, so over-lie these as to unite in every position all the colors of the spectrum, giving perfectly white light and achromatic vision of the object.

29. **Advantage of over-correcting the Object-Glass.** The eye-lens E E has its focus for red rays longer than its focus for blue rays. If the object-glass had not been over-corrected, the action of the field-lens would have caused the blue image to be formed at b, and the red image at r, but the over-correction of the object-glass has placed the blue image as much nearer to the eye-lens than the red, as is required by the difference between its foci for blue and red rays, and the curvature of the images which has been reversed by the field-lens, just equals the aberration of sphericity of the eye-lens, so that it appears to the observer free, also, from this species of error, as shown by the magnified straight image which the eye sees situated at O O''.

30. **Negative Eye-piece nearly Achromatic.** Let us now examine the action of the lenses F F and E E, a little more particularly. The pencil of rays from the centre of the object is so condensed by the object-glass A M P, that it occupies but a small space about the central portion of the field-lens F F. The border rays of this pencil, after passing the field-lens, cross each other, the red rays in the red image R' R', and the blue rays in the blue image B' B', and after thus crossing, they impinge upon the eye-lens each ray on the opposite side of the axis from what it was when refracted by the field-lens; thus the lateral chromatic aberration of the field-lens will be nearly corrected by the eye-lens, but as the eye-lens has a shorter focus than the field-lens, its chromatic aberration will be in excess.

In the same manner it might be shown that the spherical aberration of the lateral pencils will be very small, because they have been so condensed by the action of the object-glass that a pencil of light from any point in the object, occupies but a very small space in either the field-lens or eye-lens.

Examining the pencil from one extremity of the object, which, without the intervention of the field-lens, would have converged to R and B, we see that its central ray, represented by the smooth white line which passed through the optical centre of the object-glass, arrives at the field-lens uncolored, but is there divided into blue and red rays, (and other colors not represented in the figure.) The blue ray, which is refracted more strongly than the red, falls nearer the centre of the eye-lens, where its refractive power is small, while the red ray, which is feebly refracted, falls nearer the border of the eye-lens, where the curvature of the lens increases its refraction, so that it emerges from the eye-lens very nearly parallel with the blue ray from which it was separated by the field-lens. In the same manner all the red rays occupy the parts of the eye-lens where the refractive power of the lens is greater than at the points occupied by the corresponding blue rays. Thus the chromatic aberration of the field-lens is very nearly corrected by the eye-lens. This is the property of the negative eye-piece pointed out by Boscovich. The excess of chromatic aberration in the eye-lens is balanced by a small amount of over-correction in the object-glass.

31. **Use of the term Achromatic Objective.** From what has been stated, it is obvious that we mean by an achromatic object-glass, or objective, one in which the usual order of dispersion is so far reversed, that the light, after undergoing the singularly beautiful series of changes effected by the eye-piece, shall come uncolored to the eye. No specific rules can be given for producing these results. Close study of the formulæ for achromatism, and accurate calculation of the due proportions of all the parts of the instrument, are essential, but the principles must be brought to the test of repeated experiment. Nor will the experiments be of any value, unless the curves be most ac-

curately measured and worked, and the lenses centered and adjusted with a degree of precision quite inappreciable to those who are not *practically* acquainted with this kind of work.

32. **Aberration Produced by Glass Cover.** When achromatic object-glasses of considerable angular aperture are accurately tested, it is found that so perfect are the corrections, so completely are the errors of sphericity and dispersion balanced or destroyed, that if the glass has been adjusted for viewing an object uncovered, a plate of the thinnest glass or mica placed over the object, sensibly disturbs the correction, producing colored fringes and indistinctness of outline in all parts of the field, so that some new method of adjustment is required. This defect (if that should be called a defect, which arose out of improvement, or which was only rendered sensible by the great improvement given to object-glasses, as the result of Mr. Lister's discovery) was first discovered by Mr. Ross, who immediately suggested a ready means of correcting it.

The aberration referred to, which is produced upon diverging rays by a piece of flat and parallel glass, such as would be used for covering an object, will be understood by the following figure:

Let GGGG be the refracting medium, or a piece of glass covering the object O, and O P the axis of the pencil perpendicular to the flat surfaces, O T, a ray near the axis, and O T', the extreme ray of the pencil incident on the under surface of the glass: then T R, T' R' will be the direc-

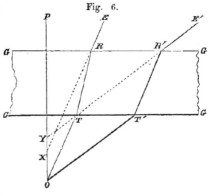

Fig. 6.

tions of the rays in the medium, and R E, R' E', those of the emergent rays.

Now if the course of these rays is continued backward, as

shown by the dotted lines, they will be found to intersect the axis at different distances, X and Y, from the surface of the glass; the distance X Y, is the aberration produced by the medium, which, as before stated, interferes with the previously balanced aberrations of the several lenses composing the object-glass. The spherical aberration thus produced by the thin glass, or other medium covering the object, being in a direction the opposite of that produced by a single convex lens, is called negative aberration. Chromatic aberration is produced by the thin glass in the same direction. The effect observed by the eye, is, that lines are not so sharply defined, and the outline of any object appears bordered with larger and thicker fringes, with colors of the secondary spectrum upon the borders of the object

33. **Objective corrected for Glass Cover.** If an object-glass constructed as shown at A M P, Fig. 5, have its anterior combination A, somewhat under-corrected, so as to leave a degree of positive aberration, and the middle and posterior combinations M P, have together an excess of negative aberration, so as to balance the under-correction of the anterior combination, then the positive aberration of this latter combination will act more powerfully upon the other two when brought as near to them as possible, and less powerfully when the distance between them is increased. When the three combinations are in close contact, their common focus, and consequently the object, is at the greatest distance from the front lens; the rays from the object are therefore diverging from a point at a greater distance than when the lenses are separated. A lens bends the rays more that diverge from a distant object, and of course its aberration is then greater, hence the anterior combination A, will have a greater positive aberration, and act more strongly upon the other combinations M and P, and this effect will vary with the distance between the anterior and other combinations.

When, therefore, the correction of the entire objective is effected for an uncovered object, with a certain distance between the anterior and middle combinations, if they are then brought into contact, the distance between the anterior combination and the object will be increased; hence the anterior combination will act more powerfully, and the entire compound objective will have a certain excess of positive aberration.

Now as a piece of flat and parallel glass placed over an object produces chromatic and spherical aberration, both negative, it is evident that it may be corrected by diminishing the distance between the anterior and middle combinations of the objective, till the positive aberration thereby produced in the object-glass balances the negative aberration caused by the medium.

This correction is essential to the performance of object-glasses of large angular aperture. Glasses of moderate angular aperture may have their corrections balanced for a medium thickness of cover, and then they will perform very well if the thickness of glass or other medium covering the object is varied within moderate limits. The mechanical arrangement of the object-glass, by which [the correction is made, to adapt it for viewing objects covered with different thicknesses of glass or fluid, is fully explained at section 58.

34. **The magnifying power of the Microscope is varied by the use of Eye-pieces of different powers,** which are designated as Nos. 1, 2 and 3, No. 1 having the least magnifying power. Nos. 1 and 2 are those generally furnished with our microscopes. No 3 is of still higher power, and is only furnished when *specially* ordered.

35. **Magnifying power varied by Draw-tube.** The power of the microscope may be greatly increased not only by using different eye-pieces and object-glasses, but also by increasing the distance between them by means of the draw-tube, with which the better instruments are supplied.

36. **Names applied to Object-Glasses.** Object-glasses are usually distinguished by the focal length of a single lens that would give the same linear magnifying power; but the distance between the object and the anterior combination of a compound achromatic objective, is far less than its designation would imply. for its optical centre, from which its true focus

should be reckoned, is at a considerable distance behind the exterior surface of the anterior lens.

37. Foci of high powers inconveniently near to the Objective. In improved triple achromatic objectives of high magnifying power and *very large* angular aperture, the diameter is only sufficient to admit the proper pencil; the convex lenses are wrought to an edge, and the concave lenses have only sufficient thickness to support their figure; consequently the entire combination is the thinnest possible, and yet the focus of the compound achromatic objective is so near the anterior surface of the front lens that an object can only be covered with the thinnest film of glass which can be obtained, when it is viewed with the highest powers.

CHAPTER II.

MECHANICAL PORTION OF THE MICROSCOPE.

38. **Modern Improvements** in the optical portion of the microscope have rendered necessary corresponding improvements in the mechanical arrangement and support of the several parts, in order to secure facility of manipulation and freedom from tremor, or vibration, which might disturb the steadiness of distinct observation.

The *microscope stand*, or *mechanical portion*, consists of the base or foot, the stage, and compound body with its support, including the mechanical arrangements for effecting the requisite adjustment of all the parts.

39. **Base.** "The instrument (says Dr Beale) should stand firmly, whether the body be inclined or arranged in a vertical position, and not the slightest lateral movement should be communicated to the body of the microscope, when the focus is altered by turning the adjustment screws. The base should be sufficiently heavy to give steadiness, and should be made either of a tripod form, or placed upon three small feet, (a simple means of insuring steadiness not attended to in many of the foreign instruments.)"

40. **Arrangement for Inclining the Instrument.** The instrument ought to be provided with an arrangement by which it may be placed in a horizontal position, or inclined at any angle, for the convenience of drawing with the camera, or measuring objects with the aid of that instrument, or to give an easy position to the observer. When the eye looks directly downwards into the microscope, the fluids collecting on the

centre of the cornea impair the distinctness of vision ; hence difficult test objects should always be viewed with the instrument inclined. An inclination of about 55° to the horizon generally affords the easiest position for protracted observation.

"It is matter of surprise, (says Dr. Carpenter, in his treatise on the microscope,) that opticians on the continent of Europe have generally neglected this very necessary convenience in the arrangement of the microscope."

41. **Stage.** The stage should be large enough to support conveniently any object, and to bring any portion of the slide, on which the object is placed, into the field of view. The stages of many European microscopes are inconveniently small. Spring or sliding clips are required to retain the object in place when the instrument is inclined. If a movable stage is attached to the microscope, the movement should be smooth and steady in every direction, and the stage should remain steady as it is placed when the instrument is inclined.

42. **Adjustment of Focus.** The mode of effecting the focal adjustment should be such as to allow a free range of two or three inches, to suit the focus of any object-glass, and there should be an arrangement for obtaining a delicate adjustment in every part of the range. To secure these objects a coarse adjustment, by rack and pinion movement, is generally employed, and a delicate fine adjustment by a screw acting upon the end of a lever.

43. **Simplicity and Facility of Adjustment** should be secured in every part, and the instrument should be so arranged in its case that it can be taken out and fitted for observation with as little labor as possible. Many interesting objects are likely to pass unnoticed, if much time is required to unpack and adjust the instrument for observation.

44. **Description of Microscope Stands.** Since we have been engaged in the manufacture of microscopes, we have carefully examined the most approved forms of English and

American microscopes, and having had the advice of several of the most eminent American microscopists, in regard to im-provements in the me-chanical arrangement of our instruments, we have devised and adopted a variety of forms and sizes of mi-croscope stands, adapt-ed to the use of every class of observers, and suited to the means of purchasers.

For steadiness of sup-port, freedom from tre-mor, convenience and simplicity of arrange-ment, and finished workmanship, we are confident that our mi-croscopes will prove satisfactory to all who may use them.

45. **Educational Microscope.** This in-strument is mounted on a firm tripod, with up-rights of japanned cast-iron. A solid limb of japanned cast-iron supports the stage and the body of the instrument, and being attached to the uprights by a trunnion joint, it allows the instrument to be in-clined at any angle. The body of the instrument slides easily and steadily in a firm, but elastic brass cylinder, attached to the japanned limb, by which means it is readily adjusted to any desired focus. The stage is two by three inches, having spring clips to retain the object in place when the microscope

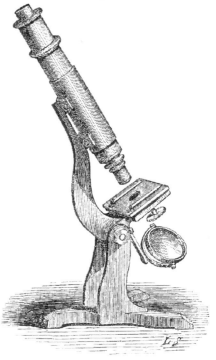

Fig. 7.

NO. 1. EDUCATIONAL MICROSCOPE.
Twelve inches high when arranged for use.

is inclined. A fine screw, with a milled head, at the right of the stage, gives a fine adjustment to the focus.

Below the stage is a diaphragm plate, with orifices of different sizes to regulate the illumination, and a space between the largest and smallest orifices to exclude all the light, and give a dark background for viewing opaque objects.

A concave mirror an inch and a half in diameter, suspended by a cradle joint, and movable in every direction, is used for illuminating the object. The mirror is so attached to the axis of the instrument, by a movable arm, that it can be turned so as to give very oblique light.

This instrument is generally supplied with two eye-pieces, and with one inch and one quarter inch objectives of second quality, giving four magnifying powers, varying from 40 to 350 diameters.

This microscope is designed, as its name implies, for educational purposes, for schools, private families, and for young people generally. Farmers, mechanics and merchants, who desire to devote some of their leisure hours to intellectual improvement, or to the investigation of those branches of natural science more or less connected with their several avocations, will find this at once a cheap, substantial and efficient microscope.

As it is very steady and delicate in its adjustments, and can be used with the higher powers, the man of science will often find it a convenient substitute for the larger microscopes, when a more portable instrument is required for special purposes.

46. The **Student's Microscope**, shown in figure 8, is mounted on a firm tripod base, with uprights of japanned cast-iron, and moves freely on trunnions, so that it may be inclined at any angle. It has a plain stage three by four inches, furnished with spring clips, which retain the object on the stage when the instrument is inclined, and yet allow it to be moved freely in any direction. Plane and concave mirrors are set in the same frame, and can be rotated in any direction, or moved upward and downward on a sliding support. The adjustment o

focus is performed by a fine rack and pinion movement, by turning either or both of the milled heads at the back of the stage.

The body of the instrument is attached to a strong arm by a bayonet joint, and can be easily removed to allow of packing into a small case. When the body is thus removed, it can be used as a dissecting microscope by inserting into the arm lenses adapted to that purpose.

This instrument is usually supplied with two eye-pieces, and with 1 inch and $\frac{1}{4}$ inch objectives of second quality, giving four magnifying powers, varying from 40 to 350 diameters. Other object-glasses and accessory apparatus can be furnished, to suit the purchaers.

Fig. 8.

NO. 2. STUDENT'S MICROSCOPE.
Twelve inches high when arranged for use.

This instrument is well adapted for the use of students in botany and natural history, and also for schools and private families, who take an interest in the wonders revealed by the microscope.

47. **Another form of Student's Microscope** is shown at figure 9, mounted on a tripod base like No. 2, figure 8, with a trunnion joint to incline it at any angle. It has plane and concave mirrors, mounted in the same manner as in No. 2. The body of this microscope slides smoothly and easily in a strong cylindrical support. A screw with a milled head,

Fig. 9.

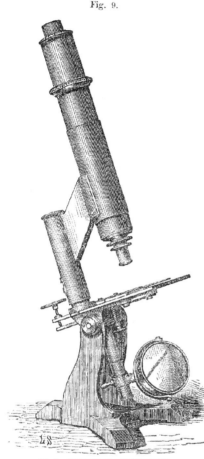

NO. 8. STUDENT'S MICROSCOPE.
Thirteen inches high when arranged for use.

seen behind the stage, acts upon a lever, which effects a delicate fine adjustment of the focus. This movement adapts this form of the student's microscope to be used with the higher powers. The stage is three by four inches, and has spring clips for securing the object. A circular plate attached beneath the stage, and carefully centered, affords attachment to the polarizer and achromatic condenser, and any other apparatus which requires to be attached beneath the stage.

Every kind of accessory apparatus may be used with this instrument. Medical students, and all others who cannot afford the larger instruments, will find this microscope well adapted to almost every kind of observations, even with the highest powers.

48. **The Student's Larger Microscope**, shown in figure 10, is made of the dimensions agreed upon by microscopists as most convenient for general use, and fitted for the application of accessory apparatus of such dimensions as to secure their greatest desirable efficiency. The tripod base is large and

strong, made of japanned cast-iron, giving firm support and freedom from tremor. The coarse adjustment is performed by

Fig. 10.

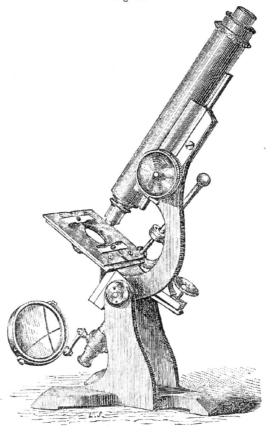

NO. 4. STUDENT'S LARGER MICROSCOPE.
Fifteen inches high when arranged for use.

a rack and pinion, by turning a large milled head, conveniently placed; the body moving steadily in a long grooved support, and being retained in any position by springs. The fine adjustment of the focus is performed by a screw acting upon a lever, which gives to the stage a delicate upward movement. Two sliding clips retain the object on the stage.

The stage, which is three by four inches, is so constructed that it can be moved smoothly and steadily in every direction, the object appearing to follow the motions of the hand upon the lever. This movement of the stage gives great facility for tracing every part of the slide in the search for delicate objects, and enables the observer to follow with ease the motions of living animalculæ, even with high powers. Beneath the stage is a circular plate carefully centered and adapted for receiving accessory apparatus. The mirrors, plane and concave, (the latter two inches in diameter,) are so mounted as to have a free and steady motion in every direction. By means of an arm, the mirrors can be thrown far out from the axis of the microscope, so as to give *very oblique* light for illuminating the object.

This instrument has a graduated *draw-tube*, by which the distance between the objective and eye-piece can be considerably increased. This increased length produces a proportional increase of the magnifying power, and thus often greatly aids in ascertaining the value of micrometer graduations. (See 71, 72.)*

49. **Stage Movable by Rack and Screw.** American microscopists generally prefer our form of stage, movable by a lever. The instruments which we keep on hand are, therefore, usually furnished with this form of stage. But we are accustomed to make to order a stage movable in two rectangular directions, by rack and screw.

---

* This instrument is often made with a plain stage, which considerably reduces the expense. It can also have added the revolving motion of the stage, as shown in the next figure. Baileys' Indicator Stage can be applied to this instrument, if desired, instead of the stage here shown. The use of cast-iron, for the base and arm of the preceding instruments, has been adopted to bring the prices within the most reasonable limits. This arrangement does not, in any manner, diminish the efficiency or beauty of the instruments. The parts made of iron are carefully smoothed and neatly japanned, and give a pleasing contrast of color with the other parts, which are of brass. But when specially ordered the base and arm are also made of brass, at a reasonable addition to the price.

---

50. **The form of Student's Microscope,** shown in figure 11, has the same tripod base, mirrors, and graduated draw-tube,

Fig. 11.

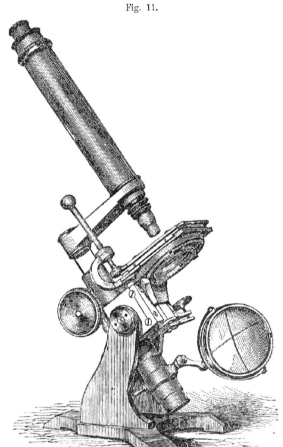

No. 5. Fifteen inches high when arranged for use.

as the preceding, and the same lever movement of the stage. In addition, the stage revolves around a steady centre, which invariably coincides with the axis of the body. Beneath the stage is a plate. carefully centered, and adapted for receiving achromatic condenser, and other apparatus. A strong triangular bar, moved by a rack and pinion, by two large milled heads, gives the quick motion of the body, while the delicate fine adjustment is effected by a screw on the left of the instrument, acting upon a lever, which gives a slow movement to the stage, as in the instrument last described.

The arm which carries the body can be turned away from over the stage; this is a great convenience in changing the object-glasses, or in performing any manipulation on the object upon the stage.

The body of the instrument, which is attached by a bayonet joint to the arm that supports it, can be easily removed, for packing in a small case, and the instrument can be readjusted for use with great facility. When the body is removed, this microscope can be used for dissecting, by inserting a single lens into the end of the movable arm in place of the body.

Chevalier's Prismatic Body, shown in the next figure, can be attached to this instrument.

51. **Chevalier's Prismatic Body,** shown in Fig. 12, as it is attached to the microscope, consists of a tube with an elbow in which is inserted a rectangular reflecting prism. The light from the object-glass enters the prism perpendicular to the first surface, and falling upon the second surface at an angle of $45°$ suffers total reflection, and emerges perpendicular to the third surface of the prism, which makes a right angle with the first. This addition is particularly useful in examining living animalcules and other objects in water, and in all other cases where the horizontal position of the stage is either necessary or desirable; and when otherwise the continued posture of the observer, in looking down vertically, would be attended with great fatigue to the eye.

This apparatus can be applied to microscopes Nos. 5 and 6. A short elbow tube, (containing a rectangular prism,) one end to be inserted in the draw-tube and the other to hold the eye-piece, is furnished to order with any of our microscopes. It is used for the same purpose as the prismatic body.

Fig. 12.

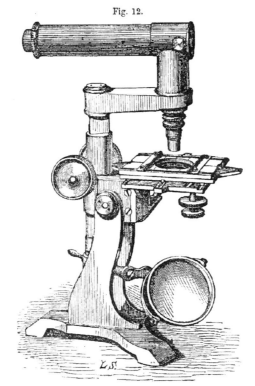

CHEVALIER'S PRISMATIC BODY, AND BAILEY'S INDICATOR STAGE.

52. **Prof. Bailey's Indicator Stage,** shown also in Fig. 12, as attached to microscope No. 5, has a graduated movement in two directions, by turning the milled heads seen under the stage. These milled heads are attached to two pinions, one revolving within the other, and they move the stage in two directions respectively at right angles to each other. On each border of the stage are scales graduated to $\frac{1}{50}$ of an inch, which serve to determine the exact position of the stage when an object is in the centre of the field. If a slide is placed in a certain fixed position on this stage, the exact position of any object seen in

the centre of the field can be determined by the graduation and recorded on the slide. The same object can then be placed in the centre of the field of view, at any future time upon any microscope having a similar indicator.

When it is considered that with the higher powers, an object measuring only $\frac{1}{100}$ or $\frac{1}{50}$ of an inch, fills the entire field of view, it will be evident that objects of great interest are brought into the field of view a second time with difficulty, and among a number of similar objects, any one which is peculiar can only be re-discovered after much patient research. But with this indicator stage, if the position of an object has been once recorded, it can be found again with great ease and certainty. The use of microscopes furnished with this stage affords great facility for interchanging slides, with the certainty of finding objects of interest singled out by correspondents.

53. **The Portable Microscope,** shown in Fig. 13, is of the same size as No. 5, but more portable.

The strong brass legs on which it is mounted, are made to fold together, and the compound body, which is attached by a bayonet joint, like No. 5, can be easily detached, and the instrument packed in a very small case.

The arm d, which carries the compound body, can be turned away from over the stage. A stout triangular bar, carrying the arm d, is moved up and down by rack and pinion connected with two milled heads, one of which is marked c. Thus either hand may be used for the quick motion of the body, which is effected with great steadiness and freedom from tremor. A delicate fine adjustment of focus is obtained by turning the milled head a, of a screw which acts upon a lever concealed in the arm d; this lever acts upon a short tube, carrying the object-glass, and sliding easily but firmly in the lower end of the body. The action of the lever upon this tube is counteracted by a spiral spring, so that an extremely sharp and delicate motion is obtained.

The elasticity of the spring diminishes the danger of injuring the object-glass, if at any time it is allowed to touch the object. Great care should be taken that no such accident should ever happen.

Fig. 13.

NO. 6. PORTABLE MICROSCOPE.
This Microscope is fifteen inches high when arranged for use.

The stage, which is three inches square and provided with sliding clips, is moved freely in every direction by means of the lever *b ;* it also revolves around a steady centre coinciding with the axis of the compound body, and a circular plate beneath the stage is carefully centered and fitted to receive the achromatic condenser and polarizing apparatus.

The plane and concave mirrors (the latter $2\frac{1}{2}$ inches in diameter) can be turned freely in any direction, or so adjusted as to give very oblique light. The instrument has a graduated draw-tube, and the whole microscope, with all the accessory apparatus, is packed in a flat mahogany case of very convenient dimensions.

54. **The Microscope** shown in Fig. 14, is of the largest class, complete in all its parts, and constructed upon the most perfect model, suggested by the combined experience of the most eminent American microscopists.

This microscope is mounted upon a strong brass tripod, with uprights of bell-metal supporting the axis, upon which the instrument can be inclined at any angle. The stage is four inches square, movable freely in every direction by a lever; it also revolves around a steady centre, coinciding with the optical axis of the microscope.

The under side of the stage is fitted for the attachment of accessory apparatus. The mirrors, plane and concave, (the latter three inches in diameter,) are mounted with a cradle joint, and movable arm attached to a sliding support, giving facility of movement in every direction.

The compound body has a graduated draw-tube, and is attached to a heavy socket which is moved up and down on a strong triangular bar of bell-metal, by rack and pinion, by which means the coarse adjustment is effected with great steadiness and freedom from tremor.

The fine adjustment of focus is effected by a screw acting upon a lever, which moves a short tube carrying the object-glass. This tube is held down by a spring, which, like a similar spring in No. 6, diminishes the danger of injuring the object-glass.

This instrument is finished in the most perfect manner throughout, and is designed for scientific institutions, and for all who desire to possess the most perfect and completely furnished instruments.

Fig. 14.

NO. 7. LARGE MICROSCOPE.
Seventeen inches high when arranged for use.

CATALOGUE OF ACHROMATIC MICROSCOPES.

55. Chemical or Inverted Microscopes. The chemical microscopist frequently has occasion to perform manipulations which are rendered difficult by the close proximity of the object-glass to the object, or to apply heat, or chemical reagents, the fumes of which might injure the lenses. To avoid these difficulties, Prof. J. Lawrence Smith has invented the inverted microscope, which he presented to the *Societé de Biologie* of Paris, in 1850.

In this microscope the object-glass is placed below the stage, and the arrangement of the several parts are such that the eye can observe the object, and, almost, at the same time, with ease, guide the hands in performing any required manipulation on the stage.

The optical arrangement of the inverted microscope is shown in Fig. 15.

Fig 15.

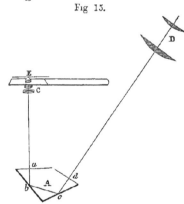

E represents the stage with the object upon it, C is the object-glass placed below the stage, A is a prism so constructed that the light entering it at right angles to the face *a*, after undergoing total reflection at *b* and *c*, emerges at right angles to the face *d*, and is viewed by means of the eye-piece placed at D. The prism A, which is the most important part of the instrument of this form, has its angles *a b*, *b c*, *c d*, and *d a*, respectively 55°, 107½°, 52½°, and 145°, consequently the body of the microscope is inclined to the perpendicular 35°, which is found to be the most convenient position for steady and protracted observation. The illuminating apparatus for the inverted microscope is placed above the stage, and consists of a reflecting prism and condensing lenses.

We are authorized by Prof. Smith, as the only manufacturers of these instruments in this country.

56. The simple form of the Inverted Microscope, shown in Fig. 16, is designed to meet the ordinary requirements of the chemist.

Fig 16.

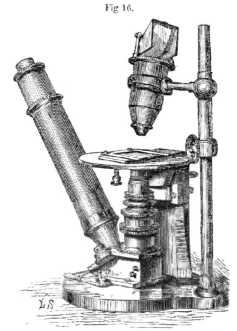

NO. 8. Prof. J. L. SMITH'S INVERTED MICROSCOPE, (Simp'e Form.)

This instrument has coarse and fine adjustments of focus, and a round stage 3¾ inches in diameter, with a glass plate for its upper surface, and spring clips to retain the object on the stage. It will be seen by the figure that the rectangular prism, used as a reflector, and the condenser, are so mounted as to be moved up and down the support, or inclined at any angle, so as to illuminate the object with very oblique light. The tube, which forms the body and carries the eye-piece, is attached by a bayonet joint, and can be removed to allow of packing in a convenient case. Polarizing apparatus and other accessories, are furnished to order with this instrument.

57. **A more complete Inverted Microscope** is mounted on a revolving foot, has coarse and fine adjustments of focus, a stage movable by a lever, and a column with rack and pinion movement to carry either the illuminator, polarizing apparatus, or oblique condenser. It can be furnished with all the accessories supplied with other instruments.

The especial use of the inverted microscope is for investigation of chemical substances, but it also affords advantages in examining all objects contained in fluids, for however deep the cell, the object lying at the bottom is seen as though mounted in the shallowest cell. By this instrument small insects, animalculæ and infusoriæ, as Desmidiæ and Diatomaceæ are viewed from their under surface, which often materially aids in investigating their structure.

58. **OBJECT-GLASSES.** Our achromatic object-glasses range from 2 inch to $\frac{1}{12}$ inch focus, with magnifying powers varying from 20 to over 1600 diameters, which may be increased, by extending the draw-tube of the microscope, to 2000 diameters. A table containing a list of our objectives, with the angular aperture and magnifying power of each, with the different eye-pieces, will be found in connection with the *Price List* at the end of this Catalogue.

To suit the requirements of different observers, we furnish two classes of achromatic objectives. Our objectives of the *First Class* have very large angular aperture, with the most perfect correction of spherical and chromatic aberration, and are mounted in the best manner, and in the most improved style, and are attached to the microscope with a bayonet joint.* The $\frac{1}{2}$, $\frac{1}{4}$, $\frac{1}{8}$, and $\frac{1}{12}$ inch objectives of this class are furnished with Mr. Wenham's form of adjustment for correcting the aberration produced by

---

* Our objectives will be attached to the microscope by a screw, for those who prefer that mode of attachment.

the thin glass which covers the object. The one inch and two inch objectives do not require this adjustment.

Figure 17 shows the method of mounting the best object-glasses, with Mr. Wenham's adjustment for the thickness of glass cover, A being a longitudinal section, and B an exterior view of the same.

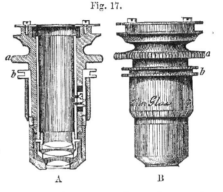

Fig. 17.

The anterior compound lens is permanently fixed to the outer tube, or that part of the mounting which connects the objective to the body of the microscope, while the middle and posterior combinations are set in a tube which slides within the other. A revolving collar, with its milled edge *b*, is attached by a screw to the inner tube, the screw passing through a slide in an inclined slit in the tube which carries the anterior combination. When the collar is revolved, the screw and slide moving in the inclined slit, the inner tube, with the posterior and middle compound lenses, is made to approach or recede from the anterior combination.

The slit in the outer tube is so much inclined that a quarter of a revolution gives to the inner tube the greatest extent of movement that is ever required. This movement of the middle and posterior combinations effects the adjustment required by covering the object with glass of any required thickness. The point on the collar marked "*uncovered*" is the point of adjustment which gives the most perfect definition of an uncovered object; the part marked "*thin glass*" indicates the correction for glass of medium thickness. Between these two points are ten or more divisions, and several equal spaces are marked off beyond this point.

This graduation is exceedingly convenient for many purposes, especially to facilitate the adjustment of the draw-tube, when using the micrometer.

59. Rules for Adjusting Object-Glasses. As many persons who use the miscroscope find some embarrassment in accurately adjusting the object-glass for the thickness of the cover on different objects, and hence fail to appreciate the real excellence of the instrument, we subjoin a few simple rules, by means of which any one may soon learn to make this correction.

Select any dark speck or opaque portion of the object, and bring the outline into perfect focus; then lay the finger on the milled head of the fine motion, and move it briskly backwards and forwards in both directions from the first position. Observe the expansion of the dark outline of the object, both when within and when without the focus. If the greater expansion, or coma, is when the object is *within* the focus, or nearest to the objective, the lenses must be placed farther apart, or towards the mark "*uncovered.*" If the greater coma is when the object is *without* the focus, or farthest from the objective, the lenses must be brought closer together, or towards the mark "*thin glass.*" When the object-glass is in proper adjustment, the expansion of the outline is exactly the same both within and without the focus.

A different indication, however, is afforded by such *test objects* as present (like the Podura scale and Diatomaceæ) a set of distinct dots or other markings. For if the dots have a tendency to run into lines when the object is placed *within* the focus, the glasses must be brought closer together; on the contrary, if the lines appear when the object is *without* the focal point, the combinations of the object-glass must be further separated.

60. Delicacy of the Adjustment for Thin Covers. When the angle of aperture is very wide, the difference in the aspect of any severe test, under different adjustments, becomes at once evident; markings which are very distinct when the correction has been exactly made, disappearing almost instantaneously when the graduated collar has been turned through no more than half a division of its graduated scale. With this form of adjustment, the transition from good to bad definition takes place with such a slight movement of the collar, that with a little practice, the best point of definition is readily found.

It often happens that the amount of balsam, or other medium, covering different objects on the same slide, varies, therefore, the adjustment should be examined for each portion of the slide, where we wish to make the most accurate observations.

For object glasses of the largest angular aperture, very great care and patience are often required in effecting the most perfect adjustment. With glasses of moderate aperture, if the adjustment has been well made for glass cover of medium thickness, the same correction will answer very well for ordinary observations, even should the thickness of the glass cover be slightly varied. For common investigations, therefore, much time may be gained by assorting the thin glass into parcels of nearly uniform thickness, and having obtained a medium adjustment for one variety, so long as the same parcel of thin glass is used, the labor of adjusting for thickness of glass cover may be dispensed with. But when glass of another thickness is employed, or when a more delicate object is to be examined, or an objective of very large aperture is employed, the adjustment for thickness of cover must be examined and made as accurate as possible.

Increasing the distance between the objective and eye-piece, by extending the draw-tube, causes disturbance of the balance of aberrations, and requires renewed adjustment by moving the graduated collar, in the same manner as for different thicknesses of glass cover.

When an object-glass is examined that has no correction for thickness of glass cover, it will usually be found that it has but very moderate angular aperture. Such is the case with almost all object-glasses made on the continent of Europe.

61. **Our Second Class Objectives** have a somewhat smaller angular aperture, and are mounted in a simpler style. They

are attached by a screw to an adapter, which is inserted by a
bayonet joint into the body of the microscope. They are very
carefully corrected for chromatic and spherical aberration, and
are adjusted for a glass cover $\frac{1}{100}$ of an inch thick. With this
correction, which is permanently fixed, this class of objectives
will perform very nearly as well if the thickness of glass covering an object is one-eightieth or one hundred and twentieth of
an inch in thickness. These objectives will resolve nearly all
the test objects mentioned in works on the microscope. The
$\frac{1}{8}$ inch of this class, with eye-piece No. 2, with good illumination, will show very clearly all the lines and dots on the *Pleurosigma angulata*. The $\frac{1}{4}$ inch will resolve the *Navicula
Baltica* and *Hippocampus* in balsam, and will resolve the
Pleurosigma angulata when mounted dry. All this class of
objectives answer well for investigations in botany, entomology,
and pathology. The low price at which this class of objectives is furnished, will commend them to all who desire cheap,
but *good* object-glasses.

CHAPTER III.

ACCESSORY APPARATUS.

62. **Micrometers.** In examining objects with the microscope, it is often desirable to ascertain their exact dimensions. Measuring instruments for this purpose, called micrometers, are sometimes applied to the object itself, but more frequently to the magnified image. Both classes of micrometers have their uses, for which they are specially adapted, and the purchaser can select that form which is best suited to the investigations for which it is to be used.

63. **Glass Stage Micrometers** are furnished mounted in brass, with lines ruled from $\frac{1}{100}$ to $\frac{1}{1000}$ of an inch. This instrument is chiefly used for determining the value of measurements made by the various eye-piece micrometers.

64. **The Cobweb Micrometer** was invented by Ramsden for telescopes, but it is equally applicable to the microscope, and, until lately, has been considered superior to all others. Ramsden employed the positive eye-piece, but the negative eye-piece is equally available, and for use with the microscope it is to be preferred.

The Cobweb micrometer is shown in Figs. 18 and 19, and consists of a negative eye-piece, in the focus of which two cobweb threads are stretched across the field; one of these threads can be separated from the other by a screw having about fifty threads to the inch. The head of this screw, shown at a, Fig. 18, is divided into one hundred parts.

Fig. 18.

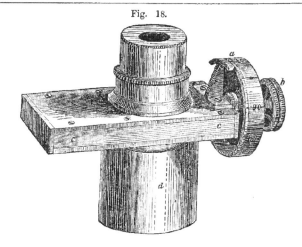

Fig. 19.

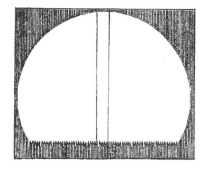

A portion of the field of view on one side is cut off at right angles to the cobweb threads, Fig. 19, by a scale formed of a thin plate of brass having notches at its edge, whose distance corresponds to that of the threads of the screw, every fifth notch being made deeper than the rest for the sake of ready enumeration. As each notch corresponds to one turn of the screw, the number of turns can be read off in the field of the instrument, and the fraction of a turn on the graduated head.

By this simple contrivance, the distance of the threads can be ascertained to the hundredth of a turn of the screw, and as the screw has fifty threads to an inch, it follows that the magnified image of an object may be measured to the five-thousandth of an inch.

With an object-glass of one-eighth of an inch focus, the image, formed at the focus of the negative eye-piece, where the micrometer threads are placed, will be magnified about fifty diameters without the power of the eye-lens; it follows, therefore, that a quantity as small as the two-hundred-and-fifty-thousandth of an inch should be appreciable by such an instrument, but in practice this has been found impossible, as no achromatic power has yet been made, capable of separating lines so close as the one-hundred-thousandth of an inch. Great care is requisite in ascertaining the value of the measurements made by this micrometer, to avoid several species of error which will be pointed out in section 71. The measurements by this micrometer are not as delicate as they appear to be, but are the most reliable that can be obtained by any eye-piece micrometer.

65. **Ross's Eye-Piece Micrometer** consists of a circle of thin glass, ruled with micrometer lines, set in a brass ring, and screwed into the lower end of the positive eye-piece. so as to be seen exactly in its focus. To adjust the focus to suit different eyes, the ring may be screwed up a little nearer to the front lens, or adjusted a little more distant. This is a very convenient form of micrometer for use with low powers.

66. **Jackson's Micrometer.** Mr. George Jackson, in 1840, invented another form of micrometer, which, with the improvements suggested by experience, he thus describes:

"Short bold lines are ruled on a piece of glass; and, to facilitate counting, the fifth is drawn longer, and the tenth still longer, as in the common rule. Very finely levigated plumbago is rubbed into the lines to render them visible, and they are covered with a piece of thin glass, cemented by Canada balsam, to secure the plumbago from being rubbed out."

The slip of glass thus prepared is placed in a thin brass frame, as shown in Fig. 20, so that it may slide freely, and is acted on at one end by a pushing screw, and at the other by a slight spring. This is inserted in the focus of the eye-lens in the negative eye-piece, through slits cut in each side for the purpose. The cell of the eye glass should have a longer screw than usual to admit adjustment for different eyes. When the frame is not employed, an inner piece of tube is drawn across the slits to prevent dust from getting between the glasses.

Fig. 20.

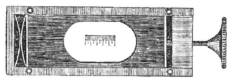

To use this micrometer, the object is brought to the centre of the field, and the coincidence between one side of it, and one of the long lines, is made with great accuracy, by means of the small pushing screw that moves the slip of glass; the divisions are then read off as easily as the inches and tenths on a common rule. The value of the divisions must, however, be accurately ascertained for each object-glass, in the same manner as for other eye-piece micrometers. The operation of measurement is then nothing more than laying a rule across the body to be measured; and it matters not whether the object be transparent or opaque, mounted or not mounted; if its edges can be distinctly seen, its diameter can be taken. By revolving the eye-piece, similar measurements can be taken across any other diameter.

A similar micrometer slide, with its pushing screw, is attached to the positive eye-piece, when this is preferred.

67. **Comparative Merits of Micrometers.** Of these micrometers, the one shown in figures 18 and 19 is the most accurate and reliable. Its measurements are all made *between* two lines, neither of which is included in the measurement, the clear space between the lines being alone reckoned. As the negative eye-piece is used, the object is clearly defined, free from all aberration. The expense of the Cobweb Micrometer alone prevents its general use.

In Mr. Ross's micrometer, the use of the positive eye-piece

somewhat impairs the definition of the object. The definition is still further impaired by the glass on which the lines are ruled, which covers the entire field of view. The lines themselves exhibit a sensible breadth, and the exact juxtaposition of the lines and object are not so easily secured, yet for low powers this micrometer answers a good purpose.

With Mr. Jackson's micrometer and the negative eye-piece, more accurate results are obtained, and although the fact that the micrometer glass covers the whole field, impairs its usefulness with high powers, yet its measurements are as accurate as are required for ordinary observations.

68. **Dr. White's Micrometer.** At the suggestion of Dr. White, of this city, we have recently made micrometers of another very simple form, which have given great satisfaction. Fig. 21 represents one of these micrometers.

Fig. 21.

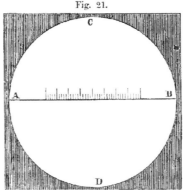

A semi-circular piece of glass, A C B, has micrometer lines ruled as shown in the figure. This is cemented to the diaphragm of the negative eye-piece, and occupies very nearly one-half of the field. The edge of the glass appears as a dark line A B, across the field, while A D B, occupying a very little more than one-half the field, is entirely unobstructed, as though no micrometer were used. This allows the eye-piece containing this micrometer to be used for ordinary observations. By

moving the object, or rotating the eye-piece, the lines can easily
be brought to measure any object in the field. The object may
be brought up to the ends of the lines without being covered
by the micrometer glass.

With this micrometer, in the eye-piece used for ordinary observation, hundreds of objects may be measured, whose magnitude would pass unnoted if it were necessary to change the eye-
piece to effect the measurement. These advantages, combined
with the very low price for which it is furnished, will, we think,
cause this micrometer to be regarded with very general favor,
and observers who have more expensive micrometers, will find
it convenient to possess this also. The lines in this micrometer
are ruled to $\frac{1}{200}$ or $\frac{1}{300}$ of an inch, to suit purchasers, but the
value of the measurements made by them must be calculated
for each object-glass, in the same manner as with other eye-
piece micrometers.

69. **Prof. J. L. Smith's Goniometer and Micrometer.**
Prof. Smith has invented a Goniometer for measuring the
angles of crystals under the microscope. It is also combined
with a micrometer. The following description of the instrument and the method of using it, are taken from the Appendix
to *Carpenter* on the *Microscope*, edited by F. G. Smith, M. D.,
American edition. Philadelphia: Blanchard & Lea. 1856.

E, Fig. 22, is the upper end of the draw-tube of the microscope with the ring k soldered to it. Over this ring k, screws

Fig. 22.

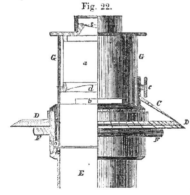

another ring F, which
serves as a support and a
centre to the graduated
circle D, which freely, but
without shaking, revolves
upon the same. Into the
bore of the ring F fits by its
lower conical end h, the
tube G, which is held in it
by the screw-ring o, that
prevents its being taken
out. Into the tube G,
which also has a free re-

volving movement, fits
the positive eye-piece a,
d being the field-lens,
and s the eye-lens.

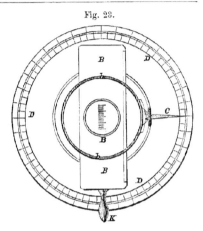

Fig. 23.

The slits $b\,b$, on opposite sides of G, (the ref-
erences in Figs. 22 and
23 are the same,) allow
the micrometer with its
mounting B B, to be in-
troduced into G, and
permit the graduated
lines to be brought into
the field of the eye-piece.
C is an index, attached
to G by the screw c, it may be taken off when the apparatus
is not used as a goniometer.

70. **Method of using the Goniometer.** Bring the object
into focus, near the centre of the field of the micrometer,
applying the finger to the knob K, revolve the micrometer till
the lines of its graduation are parallel to one side of the angle
to be measured. Revolve then, separately, the graduated
circle till zero is brought to agree with the point of the index
C. Then revolve again the micrometer by the knob K, until
the graduation lines are parallel to the other side of the angle
to be measured, when the index C will show the value of this
angle.

The micrometer lines are about $\frac{1}{200}$ of an inch apart, but
their value, when used for measurements with the different
object-glasses and eye-pieces, must be ascertained by a stage
micrometer and recorded in a table.

71. **Method of finding the value of lines in any Eye-
piece Micrometer.** For this purpose we must employ a stage
micrometer, having lines ruled at some known distance, and
this instrument should be of the very best quality, as the accuracy of all our measurements with the eye-piece micrometer
depend on the accuracy of the instrument with which their
values are determined.

To illustrate the method of making these calculations, we give the process and results in a single case, using the $\frac{1}{8}$ inch objective and negative eye-piece No. 2, in which is inserted a glass micrometer, with lines ruled about $\frac{1}{200}$ of an inch; (the exact value of these lines is of no consequence, as their value, as used for micrometer measurements, depends on the magnifying power of the glasses used.) We select a stage micrometer with lines ruled at $\frac{1}{500}$ of an inch, with no covering over the lines. We place this micrometer on the stage with the ruled lines upward. Setting the adjustment of the object-glass at the mark '*uncovered*,' and carefully adjusting the focus, we find one space on the stage micrometer covers very nearly nineteen spaces in the eye-piece micrometer; we therefore increase the magnifying power, by extending the draw-tube, till a convenient number, as twenty spaces, are covered by one space on the stage, when we find that the draw-tube is extended $\frac{7.5}{100}$ of an inch.

As now twenty spaces in the eye-piece micrometer are equal to $\frac{1}{500}$ of an inch on the stage, each division in the eye-piece measures $\frac{1}{10000}$ of an inch on the stage, when the object is *uncovered*, and the draw-tube is extended $\frac{7.5}{100}$ of an inch. We find this estimate exceedingly convenient for use, and accordingly record the conditions and estimated measurement for future reference.

But as our measurement will often, perhaps generally, be made upon objects covered with thin glass, we now place a cover of thin glass over the micrometer lines on the stage, and repeat our calculations. We correct the object-glass for thickness of glass cover, by turning the graduated collar till we obtain perfect definition of the lines on the stage micrometer, and we find. in this case, that the adjustment of· the object-glass has been turned forward twelve and one-half degrees of its graduated scale; counting the spaces in the eye-piece micrometer, covered by one space on the stage, we find, not twenty as before, but twenty-one and a fraction; we therefore diminish the magnifying power by pushing back the draw-tube till twenty spaces in the eye-piece again exactly fill one division in the stage micrometer, and we find the draw-tube extended only to $\frac{12}{100}$ inches. Thus, in covering the object with thin glass, and turning forward the adjustment of the object-glass twelve and one-half degrees, we have found it necessary to push in the draw-tube $\frac{63}{100}$ inch, or $\frac{5}{100}$ inch for every degree the adjustment of the object-glass has been turned forward.

The same method is to be pursued in estimating the value of the micrometer lines for every object-glass with which it is to be used. The values here given apply only to the particular instrument and glasses used in this calculation, for slight differences in glasses of the same name, and different lengths of the compound body, however slight, cause the measurements of the micrometer to vary. Hence every microscope should have its own table of measurements for its micrometer.

72. **Method of using the Micrometer.** Suppose we are examining a delicate Diatomacea with the instrument which we have used in making the preceding calculations, and we wish to measure the breadth of the rows of hexagonal markings which we find covering the object. We examine the adjustment of the object-glass, and find that it stands at eleven degrees, then $\frac{5}{100}$ inch for each degree would give $\frac{55}{100}$ of an inch, which the present adjustment requires to be subtracted from $\frac{75}{100}$, the distance the draw-tube is to be extended for an uncovered object, which leaves $\frac{20}{100}$ of an inch as the distance the draw-tube is to be extended to have our micrometer lines measure ten-thousandths of an inch; adjusting the draw-tube to this calculation, we see that two spaces in the micrometer cover nine rows of the delicate hexagonal markings on our object, (*navicula angulata*,) therefore the breadth of each row is $\frac{1}{15000}$ of an inch.

Suppose any object measured is equal to a certain number of spaces and a small part of an additional space, one-half, or even a quarter of a space between two micrometer lines can generally be estimated with tolerable accuracy, so that the measurements made by the glass micrometers attached to a negative eye-piece, may be relied upon for measurements as small as one twenty thousandth, or even one forty thousandth of an inch; which is a degree of accuracy sufficient for all ordinary observations. Those who desire more accurate measurements will, of course, procure the more expensive instruments. For more detailed descriptions of the method of using micrometers, we would refer our readers to the valuable works of Quecket and Carpenter on the microscope.   .

73. **Fraunhofer's Stage-Screw-Micrometer.** The labor of making the necessary calculations to determine the value of measurements made by micrometers placed in the eye-piece of the microscope, and the amount of care required to secure accurate results, have rendered it desirable to obtain a micrometer at once minutely accurate, simple in its use, and requiring no calculation of the value of its measurements.

The instrument most nearly fulfilling these requirements is Fraunhofer's *Stage-Screw-Micrometer.* The value of the measurements made by this micrometer is always independent of the magnifying power of the microscope. The instrument indicates, directly, and with great accuracy, the absolute dimensions of the object measured.

This micrometer is placed upon the stage and consists essentially of two plates, one of which, carrying the object, is moved upon the other by a micrometer-screw, having one hundred threads to an inch, with a graduated head and vernier by which motions of the plates and object upon it are accurately measured to the one hundred thousandth of an inch.

Fig. 24 shows this instrument, the upper part of the figure being a view from above, while the lower part of the figure gives a side view of the same.

B B is the body of the instrument which is fastened to the stage of the microscope by the short cylinder C, which is made elastic so as to hold the instrument securely in its place when it is inserted into the circular opening of the stage. A A is a revolving plate with spring clips to support the object to be measured. This plate can be revolved so as to bring the object in a position to measure it in any direction. This circular plate is supported on another plate in the body of the instrument, which is moved by a fine micrometer screw, with exactly one hundred threads to the American standard inch.

Fig. 24.

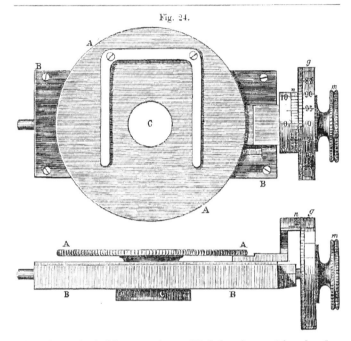

On the end of this screw is a milled head m, with a head g, having one hundred divisions; by means of the vernier n, each division of the scale g is divided into ten parts, each of which is equivalent to one thousandth part of a turn of the screw. The graduated head g can be turned without moving the milled head m or the screw to which it is attached, by which means the graduation can be set at zero in any position of the screw and object. At v is a scale showing the number of turns given to the screw, each equal to one hundredth of an inch, therefore the divisions of g are ten thousandths, and the divisions shown by the vernier are hundred thousandths of an inch, whatever may be the magnifying power of the microscope by which the object is viewed.

Accompanying this micrometer is a negative eye-piece with a cobweb drawn across the field of view, in the focus of the eye-lens, the magnitude of an object being determined by the number of turns and parts of a turn given to the screw to cause the image of the object to pass entirely across the cobweb.

For using this micrometer, the stage (if movable) is made fast by a clamp. The object-slide is placed upon the plate A A, which is rotated till the object is in a proper position to be measured. One edge of the object is brought to coincide exactly with the cobweb in the field of view, and the reading of the scale v, graduated head g, and vernier n, are accurately noted, or if the object is a small one, the graduated head g can at once be set to zero of its scale. The milled head is then turned till the object has passed entirely across the line in the eye piece, and the readings of the scale, graduated head, and vernier are again examined The difference between the first and second readings is the exact measurement of the object.

The great accuracy of this instrument, and the facility with which it can be used on any microscope, no preliminary calculations being required, will commend it to the favor of all who desire the most perfect micrometer yet invented.

INSTRUMENTS FOR DRAWING WITH THE MICROSCOPE

74. Wollaston's Camera Lucida is the instrument commonly used by artists for sketching from nature. This instrument is fitted to the eye-piece of the microscope, and enables the observer to sketch upon paper, placed on the table, the magnified image of any object seen in the microscope. It consists of a four sided prism of glass, having its faces and angles so arranged that light entering the first face of the prism is totally reflected by the second and third faces, and emerges perpendicular to the fourth face of the prism, and at right angles with its original direction.

When using this *camera*, the microscope is placed in a horizontal position, and the object appears projected upon the paper placed on the table to receive it, and it may be traced by a pencil which the eye sees at the same time as the object.

75. Nachet's Camera Lucida. This instrument consists of a triangular prism, having its three faces and angles equal.

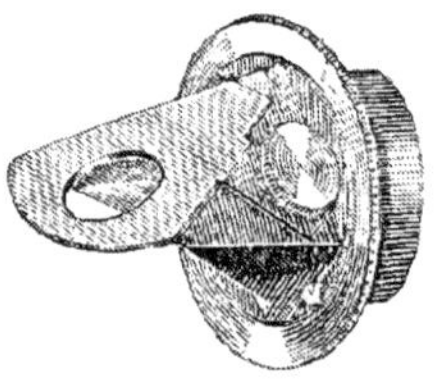

Fig. 25 shows this camera, mounted in a cap, which fits the top of the eye-piece. If the microscope is inclined at an angle of thirty degrees with the horizon, this camera may be used in the same manner as *Wollaston's Camera Lucida*, described in the previous section.

Fig. 25.

With this *camera* the microscope is sometimes placed in a perpendicular position, and the drawing made on a table which is inclined.

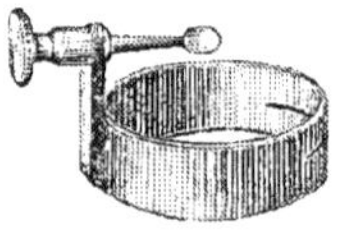

76. Soemmering's Steel Speculum. This is a plane speculum of polished steel, smaller than the ordinary pupil of the eye, commonly set at an angle of 45°, and mounted as shown in Fig. 26, and attached to the eye-piece in the same manner as the camera.

Fig. 26.

This speculum is so mounted that its angle of inclination may be changed to project the image on any part of the paper where it is most convenient. With this instrument, the reflected image of the object, and the pencil, both appear together on the paper, and the microscope may be placed at any angle which is found most convenient.

77. Using the Camera. In using the camera lucida, or speculum, some care is required to see both the object and the pencil at the same time. Sometimes the light on the paper is too strong, and it requires to be shaded by a book or other object. In the evening, on the other hand, the pencil and paper require additional illumination. If the object and pencil are not both visible on the paper at the same time, the difficulty will generally be overcome by moving the eye about till the proper position is found, and when once obtained the eye should be kept steadily in that position till the drawing is completed.

To have different drawings on a uniform scale, it is necessary to have the microscope always inclined at the same angle, that is, to have the camera at a uniform distance above the paper. Nine inches will generally be found the proper dis-

tance for easy and accurate drawing. The size of the picture
will vary with the distance of the paper from the instrument.

Experience is of course required to give steadiness to the
hand and to secure accurate drawings, but it is often so desirable to make accurate drawings of objects seen with the microscope, that every one ought to practice till he can draw with
facility. It makes little difference which of the three instruments described is used, a person draws best with the instrument to which he has become accustomed.

78. **Camera Lucida applied to Micrometry.** If the
microscope is placed at an angle convenient for using the
camera, and care is taken to adjust it always in the same position, the lines on a stage micrometer may be projected on
paper, by means of the camera, and form an accurate scale
for measuring any object drawn with the same power, and
with the instrument in the same position. Supposing the
divisions of a stage micrometer, whose real values are $\frac{1}{200}$ of
an inch, are projected with such a magnifying power as to
be at the distance of one inch from each other on the paper,
it is obvious that if any object is delineated on the paper with
the same power, every inch of the drawing corresponds to
$\frac{1}{200}$ of an inch on the object, $\frac{1}{5}$ of an inch would equal one
thousandth of an inch in the object, and so on. We may there-
fore draw parallel lines on the paper, subdividing the spaces
formed by projecting the micrometer lines, and the scale thus
formed will serve for measuring any object we may examine

A similar scale may be prepared with each object-glass, and
by viewing any object through the camera, with the scale
placed below, we determine at once its magnitude. When
sufficient magnifying power is used, the instrument properly
adjusted, and the scale thus made is minutely divided, great
accuracy may be obtained.

79. **Movable Diaphragm-Plate.** "No microscope stage
(says Dr. Carpenter) should ever be without a diaphragm-
plate, fitted to its under surface, for the sake of restricting the
amount of light reflected from the mirror, and of limiting the
angle at which its rays impinge on the object." This apparatus, shown at Fig. 27, is attached to the under side of the

stage by a bayonet joint. The diaphragm having a variety of openings of different form and size, turns
upon a pivot so situated that each
opening is successively brought into
the axis of the microscope. The
space between the largest and smallest openings is greater than between

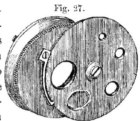

Fig. 27.

any other two, and is designed to exclude all transmitted light
and give a dark background when viewing opaque objects.

A bent spring is attached to the fixed part, and rubs against
the edge of the movable plate, which is provided with notches,
so arranged that when either of the holes is brought into its
proper position the end of the spring drops into the notch.

In the plain form of diaphragm all the openings are circular,
but in the more expensive kinds, openings of a variety of
forms are employed, some excluding the central rays, others are
crescent shaped, or semi-circular, admitting only light from one
side.

When the eye becomes fatigued by too strong light, or by
the intense glare or yellow rays of artificial light, relief is
afforded by inserting in the body of the diaphragm a piece of
gray, neutral tint, or light blue glass, by which the light can
be modified to any extent required.

80. **Bull's-Eye Condenser.** This is a large plano-convex
lens, mounted on a brass stand and pillar. The lens is attached
by a cradle joint to a revolving arm, which is supported by a
sliding tube, so cut as to act as a spring and retain the arm and
lens steady at any elevation and in any position. For illuminating opaque objects, the bull's-eye is so adjusted above the
stage, with its flat side towards the object, as to bring the light
to a focus upon the object on the stage. In using artificial light,
the large bull's-eye is to be placed with its plane surface
towards the light, and so adjusted that the beam of light
transmitted shall be about the size of the mirror, or of the
smaller bull's-eye, when the rays will be nearly parallel, or
slightly converging. The mirror, or smaller bull's-eye, may
then be used to bring the light to a focus upon the object. If the

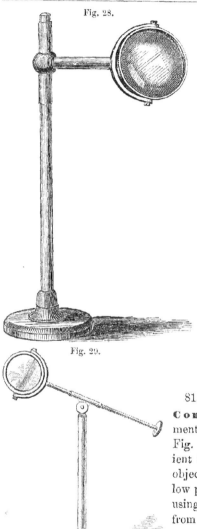

Fig. 28.

Fig. 29.

smaller bull's-eye is used to bring the light to a focus, its flat surface should be turned towards the object.

By following these directions, very little spherical aberration is produced by the form of the bull's-eye, and the illumination is rendered more effective than by any other position of the lenses.

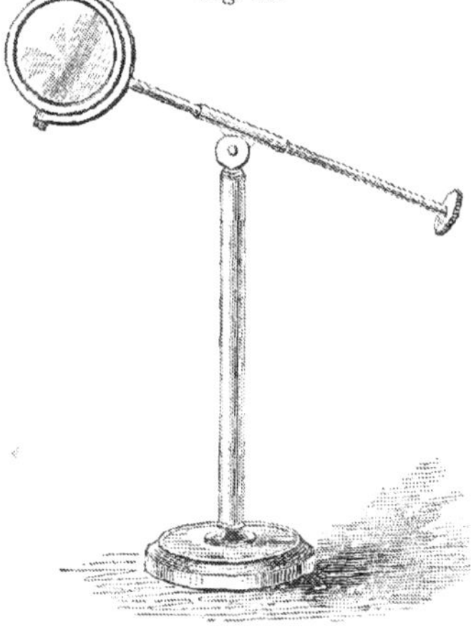

81. **Smaller Bull's-Eye Condenser.** This instrument is mounted as shown in Fig. 29. It is very convenient for illuminating opaque objects to be viewed with low powers. The method of using it will be understood from the preceding section.

82. **Achromatic Condenser.** For low powers the concave mirror, placed below the stage, furnishes all the light which is required, but for developing the best effect of the higher powers, as the $\frac{1}{4}$, $\frac{1}{8}$ and $\frac{1}{12}$ inch objectives, it is sometimes necessary to have the object illuminated with achromatic light highly concentrated. If the object-glass has an angular aperture of from ninety to one hundred and sixty degrees, and the pencil of light by which the object is illuminated is condensed at an angle of only fifty degrees, (and especially if it is also affected with spherical and chromatic aberration,) it is evident that unless the object itself disperses the light, there will be no light to be taken up by the marginal part of the object-glass. It is found that generally objects do thus disperse the light to a limited extent, but that to secure the fullest advantage from object-glasses of large angular aperture, it is necessary to illuminate the object with a pencil of achromatic light, condensed at an angle bearing a considerable proportion to the aperture of the object-glass.

This object is secured by passing the illuminating pencil through an achromatic combination of lenses, of large aperture, placed beneath the stage.

Fig. 30.

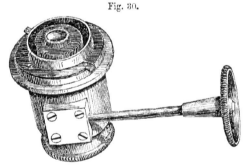

The achromatic condenser is so mounted that its focus can be easily adjusted to the exact position of the object. The apparatus in which it is mounted, is shown in Fig. 30, and is attached to the under side of the stage by a bayonet catch shown in the upper part of the figure. An inner cylinder, which carries the condenser, is moved by rack and pinion by turning the milled head, which serves to adjust the focus.

The achromatic condenser fits to the inner tube by a bayonet joint, and when it is removed, a diaphragm, or *Nachet's Oblique Prism*, may be inserted in the same brass work. The achromatic condenser furnishes a pencil of light which is, 1st, free from color; 2nd, free from spherical aberration; 3rd, condensed at a very large angle. These qualities render it exceedingly valuable for displaying delicate objects viewed with the higher powers.

To use this condenser, first place an object on the stage, illuminating it by means of the concave mirror, and with a low power bring the object into exact focus; now remove the object without altering the focus of the object-glass, attach the achromatic condenser and illuminate it by means of the plane mirror; now by turning the milled head, so adjust the condenser that the image of window bars or trees may be seen in the microscope; then turn the mirror so as to reflect light from a white cloud, or other strong light, through the condenser; then place the object on the stage and attach one of the higher powers to the microscope, and when the object is brought into focus it will be seen most beautifully illuminated, and delicate structures, scarcely distinguishable by other methods of illumination, will be clearly defined.

83. Nachet's Prism for Oblique Illumination. This prism is so contrived as to illuminate the object with a pencil of light from one side of the axis, the form of the prism being previously constructed so as to give to the light any angle of inclination required. Let M O, Fig. 31, be the axis of the microscope, b c, c d, d q, q b, are the four sides of Nachet's prism, w O v is the angle at which the pencil of light is required to fall upon the object O. To the side d q, is cemented a lens, having such a focus as to condense the light, transmitted by the prism, upon the object O, on the stage of the microscope. This prism is set in a piece of tube C C, a diaphragm placed below cuts off all the light which would not pass through the prism and be brought to a focus at O.

When this prism is used a beam of light L, is reflected

upwards by the plane mirror M, so as to enter the prism. It
then suffers total reflection in
the direction $w\,v$, and is again
reflected by the opposite force
of the prism in the direction v
O ; as it emerges from the convex surface of the lens $d\,q$, it
is condensed and brought to a
focus upon the object O, which
is in the focus of the object-
glass immediately above it.

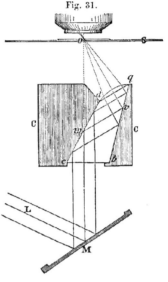

Fig. 31.

This prism is so mounted below the stage of the microscope that it can be revolved
to give light from any direction, or illuminate successively every side of an object.
This method of illumination
brings out many delicate
markings, and reveals pecu-
liarities of structure not otherwise appreciable. See Section 16.

When the microscope is furnished with an achromatic condenser, Nachet's prism can be inserted in the same brass work.

84. **Lieberkuhn Speculum.** This instrument, shown in
Fig. 32, consists of a small concave metallic reflector L L,
attached to a short tube which is fitted to the lower part of the
object-glass A. The polished surface presents that degree of
concavity which is adapted to bring to a focus, upon the object,
the beam of parallel light reflected from the plane mirror
below the stage. These little reflectors are called Lieberkuhns,
from the name of their inventor.

The rays of light reflected from the mirror B, Fig. 32, pass
through that part of the slide S S, not covered by the object,
and being again reflected by the Lieberkuhn L L, are brought
to a focus upon the object, which is then seen by reflected light.

If the object is transparent, the small stop or *dark well* D,
gives a black ground behind the object, which is then seen

Fig. 82.

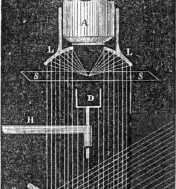

brilliantly illuminated upon a dark field. The dark well is made in the form of a cup, in order that the bottom may not be sufficiently illuminated to form a light ground to the object, which might happen if the disk were employed.

Three sizes of these dark wells are usually supplied, the largest being always used with the lower powers.

The Lieberkuhn gives a full and uniform illumination on every side of the object, while the bull's-eye gives only oblique light from one direction. The Lieberkuhn is also available to illuminate opaque objects viewed with object-glasses of such short focus as to preclude the use of the bull's-eye.

85. **The Erector** is inserted in the lower end of the draw-tube of the microscope. Its use is to reverse the position of the image, (which is inverted in the compound microscope,) so that it shall appear in the true position of the object. It is used with low powers for dissecting and other manipulations, where the hands require to be guided by looking through the microscope.

86. Among the recent valuable additions to the microscope, should be mentioned the **Orthoscopic Eye-piece**, invented by Mr. Charles Kellner, optician of Wetzlar. It is adapted to micro-scopes, and also to telescopes of all kinds, the dialytic included. It gives a large field of view, free from curvature or distortion of any kind, perspectively correct, with sharpness of definition throughout its whole extent, without the blue ring which encircles the borders of the field in the ordinary negative eye-piece.

87. **Compressor.** It is often required to tear up delicate portions of tissue upon the field of the microscope, or to float them out, as it were, from the general substance under examination, by pressing down the thin glass which covers them; many parts of plants are also better seen when slightly compressed.

Fig. 33.

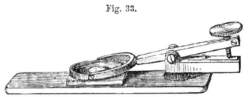

The compressor, Fig. 33, enables us to apply any amount of graduated pressure upon the thin glass which covers the object. The lever bears at one end a flat brass ring, which moves on a universal joint, and which can be elevated or depressed by turning the screw at the other end of the lever. When the screw is loosened the lever can be turned around on the pivot which secures it to the plate; the ring being thus turned away, the object on the glass plate, which covers the opening in the compressor, can be changed.

If desired, an ordinary slide can be placed upon the plate of the compressor, and the cover pressed down upon the object on the slide. Generally a plate of glass is cemented over the opening in the compressor, and a circle of thin glass is cemented to the movable ring attached to the lever. These glasses can be easily removed, if desired, or their places supplied by new ones, if they chance to be broken.

88. **Animalcule Cage with Screw.** The animalcule cage shown at Fig. 34, consists of a brass plate, or slide, carrying a short cylinder, which supports a circular plate of glass, over which fits a cap bearing a circle of *thin glass;* a screw collar retains the cap in place, and, when screwed down upon it, produces moderate

Fig. 34.

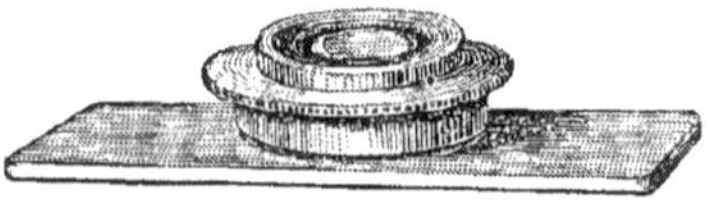

pressure sufficient to retain any animalcules in place while they are examined.

This instrument may also be used in some cases instead of the "*compressor.*"

89. **The Simple Animalcule Cage** differs from the preceding only in having the cap retained in place by a cylindrical ring, so cut as to act as a spring. The use of this instrument is the same as the preceding, though the amount of pressure obtained by it is somewhat less.

90. **Stage Forceps.** The very convenient instrument shown in Fig. 35, can be attached to the stage and made fast by turning the screw with a milled head.

These forceps slide like a pencil in a cylindrical support, while the jointed arm and pivot allow of motion in any

Fig. 35.

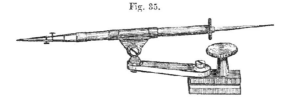

direction. Minute insects or other objects can be held by this instrument in any position required. One end of the instrument carries a needle which can be used for the same purpose.

Hand Forceps, both of brass and of steel, to be used with the microscope, are furnished to order.

91. **Frog Plate.** For viewing the circulation of the blood, the most convenient subject is the common frog. The capillary circulation in the thin transparent web of the foot, or in the tongue, affords the most interesting exhibition the microscopist can enjoy. Some care is required in so arranging the little animal as to avoid giving him pain, or *even stopping the circulation,* by undue pressure on some part of the circulatory system. The FROG PLATE is an instrument devised to secure these objects. An extra stage or brass plate, about two and a half by three inches, having a central opening, is attached to the ordinary stage of the microscope by a short piece of tube attached

to its centre, and so cut as to act as a spring when inserted in the opening of the stage. An axis extends horizontally from one side of the brass plate above described, to which is attached a plate of wood about three inches square, on which the frog is secured in a bag. This plate carrying the frog, can be rotated so as to give to the animal the most easy position for extending the foot, so as to allow the web to be spread out upon a glass plate over the opening in the stage of the microscope. On the opposite side of the brass plate are sliding sockets, in which are inserted pivots or keys like those which tighten the strings of a violin. To these pivots or keys are attached the threads which extend the toes of the frog. The threads may be tightened by turning the keys, or their position changed by moving the sliding sockets. This instrument enables one to view the circulation in the frog with great facility.

92. **Machine for Cutting Circles of Thin Glass.** The

Fig. 36.

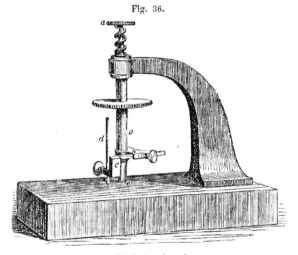

base of this instrument, which is of mahogany, supports a strong bent arm of japanned cast iron. From the end of this arm, firmly attached to it, projects downward a cylindrical guide, two inches long, designed to steady the other parts of the apparatus. Through this cylinder passes a steel stem, ending below in a conical base of brass, in the bottom of which is inserted a piece of cork or india rubber *e*, to press upon the thin glass and hold it steady. At the top is a milled head *a*, under which is a helical spring, which supports the stem, and lifts it from the glass when not in use. Outside the cylindrical guide is a tube *o*, carrying a milled head *b*, and the socket *n*, to which is attached the bent arm *c*, carrying the writing diamond *d*. The diamond is secured by a screw, and the arm which carries it can be extended at pleasure, and secured by the screw attached to the socket. The brass cone at the base of the stem supports the tube *o*, but the tube can be raised on the cylindrical guide, so that the diamond presses upon the glass only by the weight of the arm *c*, and tube *o*.

To use this instrument, set the diamond at such a distance from the axis as is required to make the circular covers of thin glass of suitable diameter, lay the thin glass to be cut upon a plate of glass previously moistened so as to make the thin glass adhere ; slide the plate carrying the thin glass under the diamond, and hold it firmly by placing a finger upon the milled head *a ;* the diamond will then rest gently upon the glass, and may be revolved by turning the milled head *b*. Removing the pressure from *a*, move the plate and thin glass to a position convenient for cutting another circle, and so continue till the piece of thin glass has been all cut into circles. When the plate is removed the circles of thin glass can be easily separated. The thin glass lies more firmly in its place, and is less liable to crack when laid upon a plate of wet glass, than if laid upon wood or dry glass. In turning the milled head *b*, care should be taken not to press too heavily upon the diamond point. A very light cut only is required, while a heavy scratch is apt to fracture the glass.

Glass covers may be cut of different sizes, so as to use up all irregular fragments of thin glass, which could not be profitably cut into squares.

Circular covers in general look better upon microscopic objects than squares, and with this instrument they may be cut with great facility and economy, as more circles can be cut from the same glass, than squares of equal breadth.

93. **Instrument for making Cells of gold size, or of other fluids.** In mounting microscopic objects very shallow cells are often required, which are most conveniently made of some fluid cement, as gold size, japan varnish, or some other similar material. This is effected by placing the slide upon a revolving table, and with a suitable brush laying on the cement evenly in a circle in the centre of the slide, as it revolves upon the table.

The instrument used for this purpose consists of a small slab of mahogany, into one end of which is fixed a pivot, whereon a circular turn-table of brass, about three inches in diameter, is made to rotate easily, a rapid motion being given to it by applying the fore-finger to a milled head below the revolving table. A circle about an inch in diameter, traced upon the centre of the table, serves to centre the glass slide which is laid upon it, and spring-clips retain the slide on the table. A camel-hair pencil, dipped in the varnish to be used, is held in the right hand and its point applied to the slide at a proper distance from the centre; the table is then rapidly rotated with the left hand and a ring of varnish of suitable breadth is easily made upon the glass. The slide is then set away to dry, when, if a thicker cell is required, another coat of varnish may be applied in the same manner.

CHAPTER IV.

POLARIZED LIGHT AND ITS APPLICATION TO THE MICRO-SCOPE.

94. **Theories of Light.** Formerly light was supposed to consist of material particles, or corpuscles, projected in all directions from luminous bodies. This theory is generally ascribed to Sir Isaac Newton, but Newton really held that in addition to the projection of luminous corpuscles, each moving particle in its flight produced vibrations in the surrounding ether, similar to the waves produced by a stone falling into the water.

Huyghens maintained, in opposition to Newton, that light consisted *solely* in vibrations of an etherial medium, originated by luminous bodies, without the onward progress of any substance whatever. With some modifications, the theory of Huyghens is generally adopted by men of science at the present time, though the honor of developing, and, to a great extent, demonstrating, the *undulatory theory of light*, is due to Dr. Young, Sir David Brewster, and other modern philosophers.

95. **Double Refraction and Polarized Light** have constituted the great battle ground of science in advancing the claims of rival theories of light. Some principal facts in regard to the polarization of light by the double refraction of Iceland spar, were known to Newton and Huyghens, but the development of the *Science of Polarized Light*, dates from the early part of the present century.

In 1808, the Royal Institute of France offered a prize for the best memoir giving the mathematical theory of double refraction in various crystals, demonstrated by experiment. This prize was awarded in 1810, to M. Malus, Lieutenant Colonel in the Imperial Guards, and member of the Egyptian Institute. In 1808, while Malus was engaged in his experiments on double refraction, and was casually observing through a double refracting prism the light of the setting sun as it came reflected at a certain angle from the windows of the Luxembourg Palace, in Paris, on slowly revolving the prism between his eye and the light, he was surprised to see a remarkable difference in the brilliancy of the two images, the most refracted gradually changing from brightness to obscurity, and the reverse at each quadrant of revolution. A phenomenon so unexpected led him to investigate its cause, and in the progress of his inquiries he discovered that every substance in nature, having a polished surface, is capable of polarizing light by reflection at a specific angle peculiar to each substance. The impetus thus given to scientific curiosity, attracted to the study of polarized light a constellation of talent almost unrivaled at any period in the history of science. So extensive have been the investigations, and so wonderful the phenomena discovered, that the subject of polarized light now justly ranks as one of the most elegant and refined branches of physical optics.

The phenomena of polarized light will here be described only so far as to enable the reader to understand their application to investigations connected with the microscope. Brewster s Optics, Pereira's Lectures, and other elaborate treatises in common use, will furnish fuller details to those who desire to pursue the subject.

96. **Polarization by reflection.** When a ray of light is reflected from a plate of glass, or the polished surface of any other substance, it can usually be reflected again from another similar surface, and it will pass freely through transparent bodies. If, however, a ray of light be reflected from a plate of glass, at an angle of 56° 45', it becomes incapable of reflection at the surface of another plate of glass in certain definite positions, but it will be completely reflected by the second plate in

other positions. It also loses the property of penetrating transparent bodies in particular positions. Light, so modified, is said to be *Polarized*.

Light may be polarized by reflection *from any polished surface*, at an angle peculiar to each substance, e. g., from glass at 56° 45′, from water at 53° 11′, rock crystal 56° 58′, diamond 68° 1′.

From a very extensive series of experiments, made to determine the maximum polarizing angles of various bodies, Sir David Brewster arrived at the following law. *The index of refraction is the tangent of the angle of polarization.* It follows, therefore, as a geometrical consequence, that *the reflected polarized ray forms a right angle with the refracted ray.* It is generally known that only a part of the light falling upon a polished surface is reflected; the remainder is either dispersed or absorbed, or, if the body is transparent, it is transmitted. Now as the differently colored rays, of which ordinary white light is composed, are not refracted equally by any transparent medium, it follows that the index of refraction, and consequently the polarizing angle, varies for the differently colored rays, therefore a ray of white light will not be *perfectly* polarized by reflection at any angle, but there is a certain range of angle within which the polarization is more or less perfect, a portion of light remaining unpolarized even at the angle of most perfect polarization.

97. **Polarization by refraction.** Certain crystallized minerals and salts have the property of polarizing light transmitted through them, owing probably to some physical peculiarity in the form or crystalline arrangement of their ultimate molecules. Such are tourmaline, calc spar, (Iceland spar,) quartz, and also many artificial salts, of which sulphate of iodine and quinine (Herapathite) is the most remarkable. The transparent brownish varieties of tourmaline furnish the most convenient polarizing plates. For this purpose the tourmaline crystal is cut into plates about $\frac{1}{30}$ of an inch in thickness and polished, the plane of section being parallel to the vertical axis of the hexagonal prism, in which form this min-

eral occurs A beam of common light, transmitted through such a plate, is almost perfectly polarized, and refuses to pass through glass and other transparent media, when it falls upon them in certain positions.

The beam of light thus polarized will pass freely through a second tourmaline plate, held in the same position as the first, as shown at Fig. 37.

Fig. 37.   Fig 38.

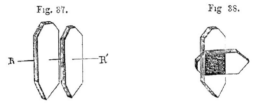

But if one of the plates is rotated before the other, and in a plane parallel with it, the light gradually diminishes till the two plates are at right angles with each other, as in Fig. 38, when the light becomes wholly obscured. As the rotation continues the beam gradually reappears, and when half a revolution has been performed the light resumes its original intensity.

This is best illustrated by looking at a candle through the tourmaline plates. If one plate only is used, the candle will be distinctly seen in every position of the tourmaline. If a second plate of tourmaline is held with its axis parallel to the axis of the first, the candle will still be seen as before, but if one plate is slowly rotated before the other, as described, the image of the candle will slowly vanish and reappear alternately at every quarter and half revolution of the plate, varying through all degrees of brightness to total, or almost total, obscurity. These changes depend obviously upon the relative position of the plates, and upon the ultimate form or physical properties of the crystalline particles of the mineral.

When the axes of the two plates are parallel, the brightness of the image is at its maximum, and when the axes of the sections cross at right angles, as in Fig. 38, the image of the candle vanishes.

A ray of light, polarized by reflection, when examined by a

plate of tourmaline, presents the same phenomena as a ray originally polarized by passing through a tourmaline plate. From its great convenience a plate of tourmaline is commonly used as a *test* of polarized light, and when so used it is called an *analyzer*.

On account of the difficulty of obtaining plates of tourmaline of sufficient size and freedom from color, other apparatus have been devised for the same purpose.

98. **Polarization by refraction through numerous plates of transparent media.** If light is transmitted obliquely through a bundle of thin transparent plates, it is polarized. Plates of the very thin glass used for covering microscopic objects, form the best polarizer of this kind. Sixteen or more of these plates, quite clean, are to be placed close together, and the bundle or package then fixed at an angle of 56° 45', to the ray to be polarized. Common window glass, or plates of mica, may be used in the same manner, but the polarization is not as complete as that produced by the action of the thin glass above referred to.

Light, which has been transmitted through such a system of thin plates, is almost entirely polarized.* It may be reflected from polished surfaces in certain positions, but not in others; it may also be transmitted through an analyzer in one position, but it is wholly intercepted when the analyzer is rotated 90°.

99. **Double Refraction.** It is well known that when light passes obliquely through water and other fluids, glass of uniform density, common salt and all cubical crystals, or crystals whose form is derived from the cube, (called *monometric* crystals,) the light is bent out of its course, but the object from which the light proceeds appears single, and in its true proportions if the transparent medium is bounded by parallel surfaces. This is called *single refraction.* Long before the discovery of polarized light, it was known that certain bodies, of which the most noted is Iceland spar, give two images of all objects seen through them in certain directions. It is now

---

* The light which remains unpolarized by this method, can only be appreciated by the most delicate analysis.

---

known that *all* crystals whose three axes are not at right angles with each other, give double images of objects seen through them in some directions. This property is called double refraction.

A great variety of substances not crystalline, possess the power of double refraction,—animal substances, such as hairs, horns, shells, bones, muscles, nerves, and other tissues; vegetable substances, like certain seeds, starch, gums, resins, essential oils and sugar in a fluid state, and many artificial substances, as glass unequally tempered. In many of these substances the separation of the two images is so slight that the double refraction is not ordinarily perceived. The peculiar character of such substances will be better understood in connection with the sections on *Partial*, and *Colored Polarization*.

The actual separation of the two images produced by double refraction, varies greatly in different substances; to show the effect, therefore, a considerable thickness of the substance is generally required: transparent crystals afford the best illustrations of this peculiar property.

100. **Double refraction of Iceland Spar.** Iceland spar, otherwise known as crystallized *carbonate of lime, calcareous spar, or calcite*, exhibits in a beautiful manner the phenomena of double refraction. It crystallizes in the form of oblique rhombic prisms, as shown in Fig. 39.

Fig. 39.

It is bounded by six equal faces, all rhombs, meeting each other at angles of 105° 5′, or its supplement 74° 55′. It has three axes, one of which, called the vertical or major axis, *a b*, Fig. 39, is equally inclined to each of its six faces, at an angle of 45° 23′. A plane *a c b d*, joining two obtuse lateral edges, is called its *principal section*. This crystal cleaves perfectly parallel to either of its six natural faces.

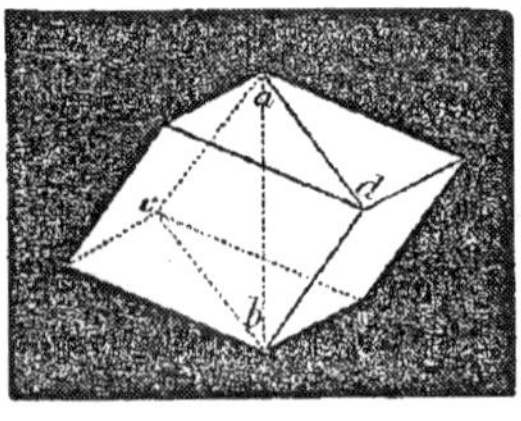

This mineral is very transparent, and in pure specimens quite colorless and free from fractures. Any object, as a line or point, seen through any of the faces of this crystal, in any direction except in a line parallel to *a b*, will appear double, owing to the subdivision of the pencil into two *beams*. As the imaginary line or axis *a b*, is that in which no doubling of the image takes place, it might with propriety be called the *axis of NO double refraction*, but as the amount of separation of the two rays depends on their position with reference to *a b*, that line is termed the *axis of double refraction*, on all sides of which double refraction takes place. This line is fixed only in direction; every other line parallel to *a b*, is equally a line of *no double* refraction. The other parts of the crystal can be split off by natural cleavage so as to leave any line parallel to *a b*, the perpendicular or major axis of the remaining crystal.

Figure 40 shows the appearance of lines A B, C D, E F, G H, and a circle drawn around their common intersection, seen through a flat prism of Iceland spar, about an inch and a

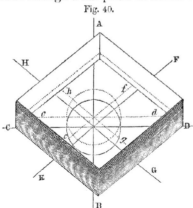

Fig. 40.

quarter thick. A B is parallel to the principal section of the prism expressed by *a c b d*, Fig. 39. The circle and lines are all seen in their true position when viewed in a direction perpendicular to the face of the prism, but just above the lines, and parallel to them, are seen with equal brightness, the dotted lines *c d*, *e f*, *g h*, and about their common intersection a second circle, the double image of the first.

The pencil of light is divided into the *ordinary* and *extraordinary* rays. The extraordinary image of a line, seen through a double image prism, is always parallel to the ordinary image. The extraordinary image of the circle shows clearly that every point is displaced in a plane parallel to A B, or the principal section joining the obtuse lateral edges of the prism. Every line drawn parallel to A B, will appear single, and every other line will appear double, when viewed respectively in a direction perpendicular to a face of the prism.

101. **Polarization produced by double refraction.** If one face of a prism of Iceland spar is covered by an opaque substance, (as paper or tin foil,) in which a small hole has been pierced, and this hole is viewed from the opposite face of the prism, held before a beam of light, two illuminated discs will be seen. As the spar is revolved before the eye, one image (*the ordinary image*) remains stationary, while the other (*or extraordinary image*) appears to revolve around the first. Examined by an analyzing plate of tourmaline, both of these images are found to be perfectly polarized. This is at once evident when the analyzer is rotated in front of the stationary prism, the two images alternately disappear and appear again at every 90° of the revolution of the tourmaline plate, one arriving at its maximum brightness when the other disappears, and vice versa, the maximum brightness of both images being equal.

If now, in place of the tourmaline, a second prism of Iceland spar is used as an analyzer, and is held with its principal section parallel to that of the first prism, both images will still be seen, but separated twice as far as when one prism only is used. If, now, however, the second prism be revolved 90°, 180°, or 270°, only one image remains. But in all other positions of the second prism, each of the images produced by the first prism is doubled, so that *four* images will be seen at the same time.

The ordinary and extraordinary rays, on issuing from a doubly refracting prism, are parallel to each other, and it is clear from the preceding observations, that they are polarized in planes at right angles to each other. A substance having such properties as we find in Iceland spar, must be of great value as a means of analysis and polarization of light.

102. **Nicol's Single Image Prism.** This beautiful contrivance, devised by Mr. Nicol of Edinburg, is of the greatest value to the microscopist of all known means of polarization and analysis, since it furnishes a perfectly colorless pencil of polarized light, and when of sufficient size allows of brilliant

illumination. This prism is constructed from an elongated rhombohedron of Iceland spar in the following manner : The natural face P of the prism, Fig. 41, which makes, with the ob-

Fig. 41.

tuse lateral edge K, an angle of about 71°, is ground away so as to form a new face, making an angle of 68° with the edge K, and a right angle with the plane of principal section joining the obtuse lateral edges K and K', which is the same as the section *a c b d*, Fig. 39. From the obtuse solid angle E, the prism is sawn through in the direction E F, making a right angle with the new terminal face, and also with the plane of principal section. From F another terminal face is constructed at right angles with the section E F and with the principal section, and making an angle of 68° with the edge K'.* All the new faces are now carefully polished, and the two parts of

Fig. 42.

the prism are cemented together, in their former position, with Canada balsam. The lateral faces of this compound prism are all painted black, leaving only the terminal faces for the transmission of light.

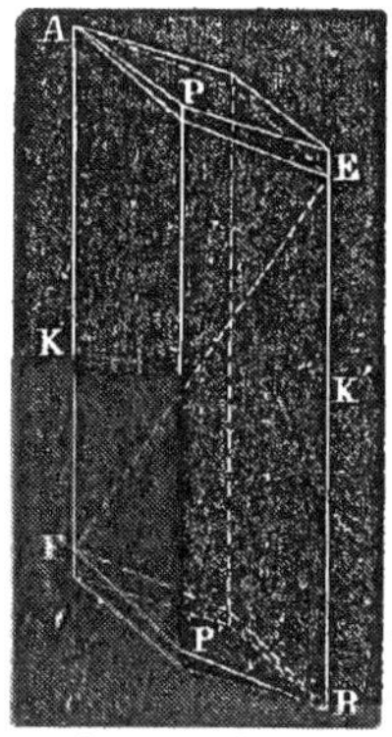

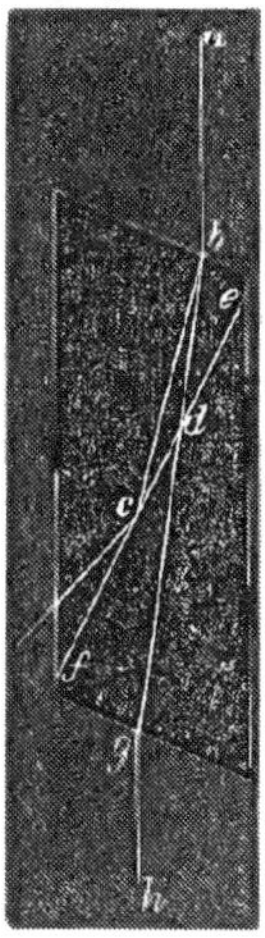

Figure 42 represents a section of Nicol's prism through the obtuse lateral edges, and shows the course of the two polarized rays into which common light is divided by this prism. A ray of common light, *a b*, entering this prism, is refracted into the ordinary ray *b c*, and the extraordinary ray *b d*. The index of refraction of Iceland spar, for the ordinary ray, being 1.6543, and that of balsam only 1.536, the ordinary ray suffers total reflection at the surface of the balsam, and cannot pass into the lower part of the prism, unless the incident

---

* In the manufacture of Nicol's prisms, the inclination of the terminal faces is sometimes varied to suit particular purposes.

ray diverges very widely from the axis of the prism. The extraordinary ray has a refractive power so low that it is not reflected by the balsam, unless it is very nearly parallel with its surface, but passes freely through it into the lower part of the prism and emerges in the direction *g h*, parallel to the incident ray.*

Nicol's prisms are capable of transmitting pencils of light, perfectly polarized, varying from 20° to 27°. The prices of these prisms vary with the size and purity of the crystals.

103. Common and Polarized Light Contrasted.

A *Ray of Common Light*,	A *Ray of Polarized Light*,
1 Is capable of *reflection* at oblique angles of incidence, in every position of the reflector.	1 Is capable of *reflection* at oblique angles of incidence, *in certain positions only* of the reflector
2 Penetrates a plate of *tourmaline* (cut parallel to the axis of the crystal) in every position of the plate	2 Penetrates a plate of *tourmaline* (cut parallel to the axis of the crystal) in certain positions of the plate, but in others it is wholly intercepted
3 Penetrates a *bundle* of parallel glass plates, in every position of the bundle.	3. Penetrates a *bundle of parallel glass plates* in certain positions of the bundle, but not in others
4 Suffers *double refraction* by Iceland spar, in every direction except that parallel to the major axis of the crystal	4 Does not suffer *double refraction* by Iceland spar in every direction, not parallel to the major axis of the crystal In some other positions it suffers single refraction only.

THEORY OF POLARIZED LIGHT

104. **Undulatory Theory.** In attempting to account for the phenomena of polarized light, the most satisfactory explanations are furnished by the undulatory theory of light, first proposed by Huyghens and more fully investigated by Dr. Thomas Young.

According to this theory, the particles of luminous bodies

---

* The refractive index of Iceland spar, for the extraordinary ray, varies by a somewhat complicated law, ranging from 1 483 to 1 651, but for such pencils of light as can be transmitted by Nicol's prism, it varies only between 1 5 and 1 56. The extraordinary ray in this prism never suffers total reflection by the balsam, unless it approaches within about 10° of its surface.

are constantly in a state of alternate contraction and expansion, and are capable of communicating their own motion to a very subtle ether pervading universal space, (and even solid bodies themselves,) through which vibrations are propagated like waves through water, but with immensely greater velocity.

105. **Illustrations of wave motion.** Let a rope, made fast at one end, be stretched in a horizontal direction, while the other end is held in the hand; if this end is moved quickly upwards and downwards at regular intervals, an undulatory motion will be propagated through every part of the rope, by a series of tremors or waves passing along from one end to the other. The vibratory motions will all take place in a perpendicular plane, the motion of each particle being at right angles to the general direction of the rope. This vibration of the rope may represent the vibrations of one of the polarized rays into which common light is separated by double refraction.

Let another similar rope be agitated in the same manner from right to left, the particles of this latter rope will vibrate in a plane perpendicular to the vibrations of the other rope, and will serve to illustrate the vibrations of the other ray of light, polarized at right angles to the first, by the action of the doubly refracting medium.

If we suppose a single rope agitated successively in an infinite number of planes, varying through every degree of the circle, the differently inclined vibrations, following each other at infinitely short (but successive) intervals, while the vibrations would take place in every possible plane, the successive waves, by which the vibrations would be propagated, would advance like the coils of a spiral, or the threads of a screw.

Let us suppose this rope, whose waves are propagated in a spiral direction, gradually restrained by the approach of two plane resisting surfaces, parallel to each other, the spiral motion would be gradually obliterated, and vibration would be continued only in a single plane, as was supposed in the case of the first rope. This may serve to illustrate, somewhat, the action of the tourmaline, which transmits light polarized or vibrating only in a single plane.

106. Polarization illustrated by resultant motions.
Every student of mechanics knows, that two forces at right
angles to each other may combine and form a resultant force
represented in direction and intensity by the diagonal of a
parallelogram, the sides of which represent the direction and intensity of the original forces; and that a single force, represented in direction and intensity by the diagonal of a parallelogram, may be resolved into two forces at right angles to each
other, which will be represented in direction and intensity by
the sides of the parallelogram.

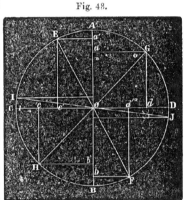

Fig. 48.

Applying these principles
to illustrate the polarization
of light, let O, Fig. 43,
represent the centre or axis
of a ray of common light
passing in a direction perpendicular to the plane of
the paper.   Let A B, G H,
D C, F E and I J, represent
transverse sections of the
planes in every direction in
which the ray of light causes
the luminiferous medium to
vibrate, we can always se-
lect two planes, as A B, C D, at right angles to each other,
which shall correspond with the planes of polarization in which
the light vibrates after double refraction.   The vibrations in all
the other planes, in which ordinary light is supposed to vibrate,
may be resolved into vibrations in the planes A B and C D.
Thus the vibration O G will be equivalent to two vibrations
represented by O a and O d; O F will be equivalent to O b
and O d'; O H will be equivalent to O b' and O c; O E will
be equivalent to O a' and O c', and so on.   W can thus resolve
the vibrations in any number of planes into others in the
planes A B and C D.

Vibrations O I, very nearly coinciding with one of the planes
C D, will give a resultant intensity in the direction of that

polarizing plane, almost equal to the original intensity, and but a feeble intensity in the other polarizing plane. But there will also be vibrations very nearly coinciding with each of the polarizing planes, so that the sum of the resulting vibrations in each polarizing plane, will be exactly equal.

The rays polarized in each plane will be represented in intensity by the sum of all the resultants in that plane.

Figure 44 is designed to represent a transverse section of a ray of common light, vibrating in an infinite number of planes, and at right angles to the direction of the ray.

Fig. 44.   Fig. 45.   Fig. 46.

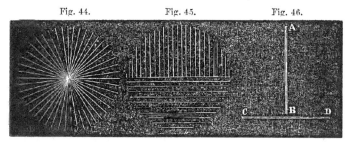

Figure 45 is designed to represent this vibrating ray, resolved into vibrations in two directions at right angles to each other, as when a ray of common light undergoes double refraction and polarization by passing through Iceland spar.

Figure 46 shows another and more common method of representing the two rays, produced by double refraction, polarized in the planes A B and C D, the fine lines in A B and C D, indicating the condensation and combination of resultant vibrations, entirely separated from the vibrations in the other plane of polarization.

107. Polarizing effect of Iceland Spar. Thus the polarizing action of Iceland spar, and all doubly refracting substances, is to separate a ray of common light, whose vibrations are in every plane passing through the direction of the ray, into two parallel polarized rays, whose vibrations are in planes at right angles to each other.

The preceding illustrations may aid in understanding that if a *polarized* ray falls upon another polarizing medium, in such a position that its vibrations are oblique to the polarizing planes or axes of the medium, they will be resolved into vibrations in *both those* axes or polarizing planes, and two new polarized rays will result, each of which might be again subdivided in the same manner by another polarizing prism placed in a position oblique to the new axes.

108. **Familiar illustrations.** A ray of common light is sometimes compared to a cylindrical rod, whereas the polarized rays are like two flat parallel rulers, one of which is laid horizontally on its broad surface, and the other horizontally on its edge. The alternate transmission and obstruction of one of the flattened beams, by the tourmaline, is similar to the facility with which a card, or flat ruler, may be passed between the wires of a cage if presented edgewise, and the impossibility of its passing in a transverse position.

We may also suppose a refracting substance, with a reflecting surface, made up of parallel fibres. Such a surface would allow the passage of all the rays in common light which vibrate in a plane parallel to the direction of its fibres and would reflect the rest; while polarized light, if vibrating in a plane parallel to the fibres, would be wholly transmitted, but if vibrating in a plane at right angles to the fibres, it would be wholly reflected.

109. **Partial Polarization.** Having already explained in the previous section how light is polarized, 1st, by reflection; 2nd, by refraction; 3rd, by transmission through bundles of thin plates; and 4th, by double refraction; it now remains to state that a great variety of substances, and in different conditions, produce *partial polarization* of light reflected from their surfaces or transmitted through them.

It is well known that *no substance either transmits, or reflects, all the light* that falls upon it; even the most transparent substances reflect a small portion of light, the proportion reflected and transmitted varying with every angle of incidence. While plate glass polarizes very nearly all the light that falls upon it at an angle of 57°, the intensity of the reflected ray is equal to only one-half the intensity of the incident ray, the other half of the light is transmitted through the glass, and is, at the same time, polarized by refraction in a plane at right angles with the plane of the reflected ray. If the light falls upon the glass at any other angle, some portion of it will be polarized by reflection, and an equal amount will be polarized by refraction, but each of the rays thus polarized will be mingled with other light that is reflected or transmitted without polarization.

If instead of transmitting light through a bundle of plates of thin glass, we transmit it through only a single plate, and at any angle, that plate will polarize a part of the light; if two plates are used, still more light will be polarized, and when a sufficient number of plates are used, all the light will be polarized.

Many substances, not perfectly homogeneous throughout, have veins, laminæ, or isolated spots, capable of producing either partial or complete polarization of the light which passes through them.

In Iceland spar the two images produced by double refraction are equal in intensity, and are both completely polarized ; but a great variety of substances give, by refraction, a secondary image so faint and so little separated from the ordinary image, that its existence is not generally recognized unless examined by some test for polarized light. There are many substances, both animal and vegetable, that possess this power of partial polarization, and which, when viewed by polarized light, exhibit an arrangement of structure totally unappreciable by ordinary light.

Even glass which has been cooled more rapidly in one part than another, or in which unequal pressure has been exerted in any manner, (as by bending or heating on one side,) is capable of polarizing some of the light transmitted through it. Many crystals found in animal fluids or vegetable tissues, possess the power of partially polarizing light. Hence, polarized light becomes the most delicate test known for discovering differences of density or structure, in animal, vegetable, or mineral substances, and is of great importance as a method of illumination in microscopic investigations.

POLARIZED LIGHT APPLIED TO THE MICROSCOPE.

110. **Polarizing apparatus.** The apparatus employed for microscopic investigations with polarized light, consist of a Nicol's prism placed below the stage and called the *polarizer* and an *analyzer*, which is usually also a Nicol's prism, set in a brass tube and inserted in the body of the microscope immediately behind the object-glass.

The *polarizer* is so mounted that it can be revolved, allowing the polarizing planes of the two prisms to be made to coincide or to make with each other any required angle. These two instruments together constitute the *Polariscope.*

Fig. 47.   Fig. 48.

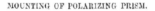

MOUNTING OF POLARIZING PRISM.   MOUNTING OF ANALYZING PRISM.

Figure 47 shows the mounting of the polarizer which is attached to the stage by a bayonet joint. The polarizing prism is revolved by turning the milled head at the bottom of the instrument.

Figure 48 shows the mounting of the analyzer, which is inserted in the lower end of the compound body, and attached by a bayonet-joint in the usual position of the object-glass, but passing upward into the body, the object-glass being attached to the lower end of the analyzer.

Figure 49 shows a section of the polarizer and the form of mounting. Above and below the Nicol's prism are circles of thin glass to protect the delicate faces of the prism from injury.

Figure 50 shows a section of the analyzer consisting of a Nicol's prism and its mounting, and protected like the polarizer by circles of thin glass.

Fig. 49.

Fig. 50.

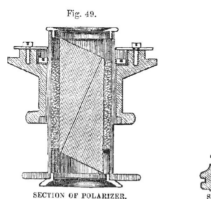

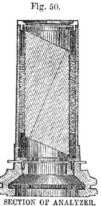

SECTION OF POLARIZER.   SECTION OF ANALYZER.

The mountings of the *polariscope* are made of different sizes, to suit both large and small microscopes, and the prices vary accordingly, depending principally upon the *size* and *quality* of the Nicol's prisms employed. The largest and best Nicol's prisms transmit the most light and are suited for the more delicate investigations.

Some opticians place the analyzer above the eye-piece, but this arrangement diminishes the extent of the field. If the analyzing prism is small, and is placed just above the object-glass, it stops out a portion of the light, but if it is of sufficient size it transmits all the light from the object-glass, and does not in any manner limit the field of view. For these reasons we prefer mounting the analyzer in the body of the microscope just above the object-glass. It may also be adapted to the lower end of the draw-tube, which arrangement allows it to be rotated by the milled ring of the tube itself.

111. **A Tourmaline plate fitted to the Eye-piece** makes an excellent analyzer, when it can be obtained of suitable size and purity of color. Such tourmaline plates are comparatively rare and costly.

112. **Herapathite or Artificial Tourmalines.** These are crystals of *disulphate of iodine and quinine*, and when of sufficient size can be used both as polarizers and analyzers. They are called Herapathite from the name of their discoverer, Dr. Herapath, but large crystals have not yet been produced in sufficient abundance to allow of their general use.

113 **Value of the Polariscope in Microscopic investigations.** When the polariscope is attached to the microscope, the polarizer being below the stage and the analyzer above it, any object can be subjected to examination with polarized light. Objects having parts of their structure more dense than others, will present greater contrast of light and shade than by ordinary light, and thus the most delicate structures, as capillary blood-vessels, nerves, cell-walls, &c , will be well defined where they could not be otherwise distinguished. Sections of horn, teeth, bones, quills, shells, and many vegetable tissues, exhibit their delicate structure under the influence of polarized light. Revolving the polarizer causes the relative brightness of different structures to vary, and is essential in developing the greatest effect of polarized light in distinguishing delicate structures Those portions of an object, possessing the power of partial or complete polarization, act like particles or veins of tourmaline to obstruct the passage of polarized light in certain positions, while the other parts of the object appear as luminous as by ordinary light. So, also, when the position of the polarizer and analyzer is such as to cut off all the light from ordinary objects, the delicate structures that possess the property of polarizing light, depolarize the light already polarized and allow it to be transmitted, showing points or veins brilliantly illuminated amid other parts of the object which appear dark.

It should therefore be laid down as a rule in microscopic investigation, says Prof. Queckett, "That every new variety of tissue should be subjected to the action of polarized light."

114. **Colored Polarization.** All substances, whether animal, vegetable, or mineral, which, by the unequal arrangement of their particles possess the property of double refraction, when placed between the polarizing and analyzing prisms exhibit *colors*, varying according to the otherwise unapprecia-

ble difference of density of their various parts, and these differences may thus be distinguished and traced out much more satisfactorily than by common light. Polarized light may be compared to a *new sense* given to the student of nature, by which he is enabled to see things wholly invisible by ordinary light.

Where the doubly-refracting properties of the tissue are too feeble to produce sufficient difference of color, the effect may be increased by placing the object over a plate of *selenite* or mica, of such a thickness as to give to the light any shade of color required.

115. The Cause of Colors produced by selenite or mica, when polarized light is transmitted through them, is easily understood by reference to the undulatory theory of light. In all doubly refracting substances, (of which selenite and mica are examples,) the ordinary and extraordinary rays move with different velocities, and consequently, when the two rays are again blended, unless the retardation amounts to a certain number of *entire* waves, the two rays will, by the interference of their waves, produce some change in the color of light. If the retardation equals any number of *entire* vibrations, the result will still be white light, the two rays conspiring to increase their mutual intensity. If one ray is retarded an odd number of half vibrations, they will mutually destroy each other and produce darkness, just as if two waves of the sea meet in such a state that the phase of elevation in one coincides with the phase of depression in another, the two will produce a level, or mutual obliteration results. Such a result in the case of light, would require the most exact adjustment of the thickness of the crystal, and would not often occur.

The interference produced by selenite and mica, are, in general, similar to the results which would be obtained by placing one prismatic spectrum upon another, in a reverse position, but not exactly superimposed upon it. The amount of overlaping would determine the resultant color.

116. Method of Varying the Colors. When a film of selenite, of uniform thickness, is placed between the polarizer and analyzer, on rotating the film a brilliant color is perceived at every quadrant of a circle, but in intermediate positions it vanishes altogether. We observe, however, that the *tint* does not change, but only varies in intensity.

If, now, the film of selenite is fixed and the polarizer is rotated, we also observe color at every quadrant of revolution which disappears in intermediate positions, but the tint changes and becomes complementary at every quadrant,—the same tint reappearing at every half-revolution. Thus, when the film alone is revolved, one color only is seen, but when the polarizer is revolved two complementary colors are seen.

The following is Sir David Brewster's table of complementary colors:

Red,	complementary,	Bluish green,
Orange,	"	Blue,
Yellow,	"	Indigo,
Green,	"	Violet reddish,
Blue,	"	Orange red,
Indigo,	"	Orange yellow,
Violet,	"	Yellow green,
Black,	"	White,
White,	"	Black.

Films of selenite, varying between .00124 and .01818 of an inch in thickness, will give every variety of tint in the solar spectrum. If two films of selenite are placed over each other, with their crystallographic axes parallel, the color produced will be that which belongs to the sum of their thicknesses. But when the two films are placed with similar axes at right angles, the resulting tint is that which belongs to the difference of their thicknesses.

A film of selenite or mica of such thickness as to produce a bright purple, or a light blue color, will be found to present the most agreeable contrast, and, as a single plate, prove most generally useful to the microscopist. Three films of selenite, which separately give three different colors, may each be mounted in Canada balsam, between slips of thin glass, and used singly, or in double, or triple combinations. As many as thirteen different tints may thus be obtained.

117. **Selenite Stage.** This instrument, which was invented by Mr. Darker, is shown in Fig. 51. It consists of a plate of brass, three or four inches long, $1\frac{1}{2}$ inches broad, and $\frac{1}{5}$ of an inch thick, having a piece of raised brass screwed to it, against which objects may rest when the body of the microscope is inclined.

Fig. 51.

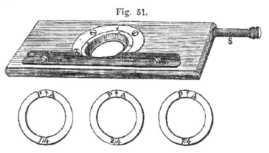

In the centre of the brass plate there is a hole, one inch in diameter, into which is fitted a ring of the same metal, with a shoulder on its under side to receive certain cells, into which plates of selenite are fitted; this ring can be revolved either to the right or the left of a central index or dart, by means of an endless screw S. P A $\frac{1}{4}$, P A $\frac{2}{4}$, P A $\frac{7}{4}$, represent three brass cells, into each of which are burnished two plates of thin glass, having between them films of selenite of different thicknesses. The dart P A, denotes the direction of the positive axis of the selenite, and the figures $\frac{1}{4}$, $\frac{2}{4}$, $\frac{7}{4}$, denote the parts of a vibration retarded by each disc, which, by their superposition and variation in position, by means of the endless screw motion, produce all the colors of the spectrum.

118. **Polarizer with Revolving Selenite Carrier.** In order to afford the greatest facility of revolving the selenite plate, and for convenience of using it, a revolving selenite carrier is attached to the polarizer, as shown in Fig. 52. The solid ring a, is attached to the stage in the usual way by a bayonet-joint. A cylinder, with a milled head c, is supported by the ring d, which revolves in the ring a. Upon d rests the selenite carrier s, covered by the cap e, so that the selenite plate is revolved by turning the milled head c. Within the cylinder

which supports the selenite, another cylinder, carrying the Nicol's prism P, is separately revolved by turning the milled

b. The Nicol's prism revolves more easily than the selenite carrier, so that the latter may remain stationary when the former is revolved. By turning the milled head c, the polarizer and selenite are revolved together. This has the same effect upon the light as if the analyzer were revolved. By means of this apparatus, the axes of the polarizing prism, selenite plate, and analyzer, may be placed in any relative

Fig. 52.

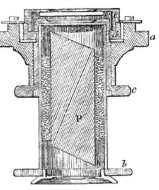

position desired, affording great facility for using polarized light colored of any tint required in microscopic investigations. We are happy to acknowledge our obligations to Dr. White of New Haven, for suggesting this improved arrangement for revolving the plate of selenite.

119. Delicate structures, viewed by colored polarized light, produce much more sensible changes upon the light than upon *plain polarized light*,—microscopic crystals are especially beautiful when viewed in this manner. The crystallization of various salts, viewed by polarized light, is a subject of great importance to the practical chemist and mineralogist. So minute a quantity as one thirteen millionth of a grain of potassa, when tested with bichloride of platinum, gives a distinct and characteristic tint, sufficient to distinguish it from every other alkali, when viewed in the microscope by this kind of light. Many substances, known in organic chemistry, are more readily distinguished by polarized light than by any other means.

The examination of microscopic structures, by polarized light, affords to the enterprising student a rich field of investigation, as yet but partially explored. As a method of investigating delicate structures it is of the highest value. The chemist may perform the most dexterous analysis ; the crystallographer may examine crystals by the nicest determination of their forms and cleavage ; the anatomist or botanist may use the dissecting knife and microscope with the most exquisite skill ; but there are still structures in the mineral, vegetable, and animal kingdoms, which defy all such modes of examination, and which yield only to the magical analysis of polarized light. A body, which is quite transparent to the eye, and which might be judged as monotonous in structure as it is in aspect, will yet exhibit, under polarized light, the most exquisite organization, and will display the result of new laws of combination which the imagination even could scarcely have conceived.

120. **List of Objects for the Polariscope.** Sufficient having now been stated to give the reader a general view of the nature and use of polarized light and its application to the microscope, we shall conclude this subject by simply giving A TABLE OF THE MORE INTERESTING OBJECTS FOR THE MICROSCOPE, WHICH ARE ESPECIALLY BEAUTIFUL WITH POLARIZED LIGHT :

ANIMAL STRUCTURES.

Bone of Cuttle-fish,	Raw Silk,
Fibres of Sponge,	Scale of Eel,
Hoof of Ass,	Scale of Sole,
Hoof of Camel,	Skin, Elephant,
Hoof of Sheep,	Skin, Crocodile,
Hoof of Horse,	Skin, Human,
Hoof of Ox,	Skin, Rhinoceros,
Horn of African Rhinoceros,	Skin, various Serpents,
transverse section,	Spicules of Gorgonia,
longitudinal section,	Whalebone,
Horn of Indian Rhinoceros,	Palate of Whelk,
Horn of Antelope,	Palate of Limpet,
Horn of Ox,	Palate of Nassa,
Horn of Sheep,	Palate of Paludina,
Polyzoaries,	Palate of Cyclostoma,
Quill of Porcupine,	Wing cases of Beetles,
Quill of Echidna,	Scales of Fishes,
Quill of Condor,	Sections of Hairs,
Tendon, Human,	Sections of Teeth,
Tendon, Ostrich,	Nerves and Muscle.
Gray Human Hair,	

VEGETABLE.

Starch, Potato,
Starch, Arrowroot,
Starch, Custard-powder,
Starch, Indian-corn,
Starch, Tous les Mois,
Gun-cotton,
Hairs from leaf of Deutzia,
Hairs from leaf of Elæagnus,
Hairs from leaf of Olive,
Hairs from leaf of Cactus,

Raw Cotton,
Raw Flax,
Siliceous Cuticle of Bamboo,
Siliceous Cuticle of Equisetum,
Siliceous Cuticle of Rice,
Siliceous Cuticle of Wheat,
Raphides,
Spiral Cells and Vessels,
Wood, longitudinal sections mount-
    ed in balsam.

MINERAL

Agate,
Arragonite,
Asbestos,
Adventurine,
Granite,
Marble,
Selenite films,

Sea sand,
Tremolite,
Satin Spar,
Sandstone,
Feldspar,
Crystals (of Titanic Iron, ?) in some
    varieties of mica.

CRYSTALS, VIZ

Acetate of Copper,
Bichromate of Potash,
Borax,
Boracic Acid,
Borate of Ammonia,
Borax and Phosphoric Acid,
Carbonate of Lime,
Chromate of Potash,
Chlorate of Potash,
Cholesterine,
Nitric Acid,
Epson Salts,
Oxalic Acid,
Oxalate of Ammonia,
Oxalate of Chromium,
Oxalate of Lime,
Oxalate of Soda,

Nitrate of Ammonia,
Nitrate of Baryta,
Nitrate of Lead,
Nitrate of Potash,
Nitrate of Soda,
Phosphate of Soda,
Salicine,
Sugar,
Sugar of Milk,
Sulphate of Cadmium,
Sulphate of Copper,
Sulphate of Magnesia,
Sulphate of Nickel,
Tartaric Acid,
Tartrate of Lime,
Uric (or Lithic) Acid,
Triple Phosphate.

CHAPTER V.

PRACTICAL DIRECTIONS.

121. Care of the Microscope. It is of the first importance that every part of the microscope should be kept perfectly clean and free from dust. For this purpose a good case is required, into which the instrument can be easily placed when not in use. In order that no unnecessary time may be lost in packing and unpacking, the upright case is more convenient, as the instrument can be placed in it nearly in the same condition as when arranged for use. Each object-glass, when not in use, should be carefully returned to the brass box provided for it, as it is thus less likely to be injured by dust or other means than when attached to the microscope. The case should be arranged with proper fittings to hold all the apparatus belonging to the microscope. Two or three drawers are required to hold slides, thin glass, dissecting needles, &c.

If the microscope gets soiled it may be wiped with fine linen or cambric. The lenses of the eye-piece may be wiped with soft buckskin which has been thoroughly freed from dust, but care should be taken that this operation may be required as seldom as possible. The object-glasses will lose their fine polish if often wiped. Some microscopists are so careful of their object-glasses that they do not require cleaning for years.

If an object-glass requires cleaning it may be brushed with a fine short camel's-hair-pencil which is used for no other purpose, or it may be moistened by breathing on it and then gently pressed with soft buckskin. When fluids are used about the microscope, care should be taken to avoid wetting the object-glasses. If this accident happens, they must be wiped dry with the buckskin.

In handling the object-glasses, care should be taken to avoid touching the surface of the lenses. The glasses should be examined when returned to the case to see if they have been soiled. Too great precaution cannot be taken to keep everything about the microscope scrupulously clean.

122. **Illumination.** For whatever purpose the microscope may be used, it is important to secure a pure and adequate illumination. The young microscopist will find daylight more easily managed, and more pleasant for the eyes, than artificial light. Even experienced microscopists can work longer and easier by good daylight, though they may have learned to manage artificial light so as to make it more available than daylight in bad weather.

The best light is that reflected from white clouds, but light from fleeting clouds is troublesome to the eyes and requires constant moving of the mirror. Light from a luminous atmosphere, near the horizon, is better than that near the zenith. The direct light of the sun cannot ordinarily be used for illuminating the microscope, but the light of the sun, reflected from white houses, or from a white plastered wall, gives a soft and beautiful illumination.

If possible, the microscope should be placed at a little distance from a window on the side of the house, opposite to where the sun is shining. If artificial light is used, the German student's lamp, constructed on the bird fountain principle, which gives excellent illumination, will be found as convenient as any other. If gas light is used, a pane of light blue glass placed between the light and the microscope, will take off the intense glare which is otherwise apt to injure the eyes. All flickering lights are very unpleasant in microscopic investigations. Observations will generally be conducted with greater ease, when the body of the microscope is inclined.

123. **Choice of a Microscope.** In choosing a microscope, reference should always be paid to the nature of the investigations for which it is to be used. The most common error among inexperienced persons, is the idea that all objects can be satisfactorily examined with a high magnifying power. This is by no means correct, for it is often as desirable to ascertain the relations of different portions of an object, as to examine minutely any single part. Small crystals, insects, and parts of flowers, hairs, &c., which are readily obtained, are best studied with magnifying powers of from twenty to one hundred diameters. Young people in schools, academies, and private families, generally first examine this class of objects, and but few such persons can devote the requisite attention to prepare specimens with sufficient care to allow of their being examined with magnifying powers higher than three hundred and fifty diameters.

The Educational Microscope, (page 25 of this Catalogue,)

has been prepared with especial reference to the greatest efficiency for this class of observers. It is also furnished at a very moderate price.

Our advice is often solicited by persons who have a limited amount of money to expend for a microscope, and who desire to know what selection would be most useful for their several purposes. For those who design to devote but a limited amount of time to the use of the microscope, we would recommend one of the smaller microscopes, No. 1, 2, or 3, with the apparatus usually attached to them as stated in our *Price List.*

To those who desire to have at once an instrument suited to the more delicate investigations at the smallest cost, we would recommend the Student's Microscope. No. 3, *furnished with First Class Objectives,* to be purchased according to their means in the following order: For the study of botany and mineralogy, the 1 inch, $\frac{1}{4}$ inch, $\frac{1}{2}$ inch, 2 inch, $\frac{1}{8}$ inch, $\frac{1}{12}$ inch objectives. For studying entomology, anatomy and pathology, the $\frac{1}{2}$ inch, $\frac{1}{4}$ inch, 1 inch, $\frac{1}{8}$ inch, 2 inch, $\frac{1}{12}$ inch object-glasses.

The Student's Larger Microscope, No. 4, furnished with 1 inch, $\frac{1}{2}$ inch, and $\frac{1}{4}$ inch objectives, Bull's-eye Condenser, Camera Lucida, and Micrometer, forms a very complete instrument with which almost all vegetable and animal tissues and fluids can be examined satisfactorily. This combination is especially recommended to Medical Students, and all others who intend to devote considerable attention to microscopic investigations, and desire to make the most economical investment of limited means, and yet secure a *first class instrument.*

124. Qualities of Object-Glasses. In considering the value of an object-glass, and its adaptation to any particular purpose, several distinct qualities are to be examined, viz: (1.) its *defining power,* or the power of giving sharpness of outline, especially on the borders of an object, or where dots or lines are examined; (2.) its *resolving power,* by means of which closely-approximated markings are distinguished; (3.) *flatness of field;* (4.) *depth of definition,* which refers to the distance above and below the focus, that parts of an object can be seen with tolerable accuracy.

1. *Defining power,* in addition to perfect workmanship, depends principally upon the *perfection of the corrections,* both for Spherical and for Chromatic aberration The difficulty of securing this perfect correction increases with the angular aperture Any inaccuracy in adjusting the object-glasses for thickness of glass cover, which every observer must arrange for himself, is more conspicuous in glasses of large angular aperture For this reason, object-glasses of moderate

aperture are more suitable to be used in schools and private families where many persons use the same microscope, and where, for want of time, in examining a variety of specimens, or from inexperience, the necessary attention cannot be devoted to adjusting the correction for thickness of glass cover.

2 *Resolving power* (correct definition being presupposed) may be said to stand in a direct relation to angular aperture, and consequently to the obliquity of the rays which can be received from the surface of an object

To measure the angular aperture of an object-glass, place the microscope, with the body horizontal, on a thin board which turns on a pivot exactly under the focus of the object-glass, set a lamp on a level with it a few yards distant, then having directed the body of the microscope so far on one side of the light that only half of the field is illuminated, leaving half of it dark trace a line by the edge of the board on which the microscope stands, now revolve the microscope horizontally to the other side of the light till only the opposite half of the field is illuminated; the angle now formed by the edge of the board and the line previously traced, is equal to the angular aperture of the object-glass　If the object-glass has a very large angular aperture, the line of demarkation between light and darkness in the field, will be indistinct, and the experiment must be performed in a dark room, with a ray of sunlight entering through a narrow perpendicular slit, by which means the exact angular aperture will be more readily determined. Without due precaution, errors of several degrees will be made in estimating the angular aperture of object-glasses　Adjusting the object-glass for a covered object, will increase the angular aperture of some object-glasses ten or fifteen degrees　Extending the draw-tube has a similar effect　Hence, in comparing the angular aperture of two objectives, they are to be examined with the same eye-piece, similar corrections for glass cover, and the same length of tube.

Some methods employed to determine angular aperture, *really* determine nothing but the angular breadth of the field of view, which is often less than the angular aperture for an object in the focus of the microscope　The real question to be determined is the angular aperture of the object-glass *as ordinarily used in the microscope;* not what is its angular aperture in other conditions where it is never placed for practical use.

3 *Flatness of Field*　To judge correctly of this quality, object-glasses should be tested with an eye-piece which gives a tolerably large field of view.　In microscopes of inferior quality, the defects of the objectives are often concealed by eye-pieces so constructed as to give but a very limited field of view

4 *Depth of Definition*　The qualities already enumerated, defining power, resolving power, and flatness of field, may all coexist in the same object-glass, but there is another quality so essential to the prosecution of microscopical researches of a certain class, and which is generally so little understood and appreciated, that we shall dwell upon it more particularly　We refer to the quality which some object-glasses possess, in addition to clear definition in the focus, of giving tolerable definition of parts a little above the exact focus　This quality may, perhaps, be called *Depth of Definition*, as it refers to the distance above or below the focus where definition ceases, and where objects, by their distance from the focus, become invisible

The value of object-glasses used for viewing tissues containing cells or vessels variously related to surrounding parts, and, in short, for the practical every-day work of the microscopist, depends very much on the quality which we have here called *depth* or *extent of definition*

As a general rule, it will be noticed that object-glasses, of the largest angular aperture, do not show parts above or below the focal plane as well as glasses of moderate aperture, and yet, in this respect, a great difference is perceived in the object-glasses of different makers.

(*a*) If two objectives, having the same magnifying power and angular aperture, are both perfectly corrected for chromatic and spherical aberration, the glass which has the greater distance between the anterior lens and the object, will have the greater *depth* of definition

(*b.*) If two objectives have the same magnifying power and the same distance between the anterior lens and the object, the glass having the smaller angular aperture will have the greater *depth* of definition, provided the angular aperture is not too small to admit a tolerable amount of light.

(*c*) It is possible, however, that an object-glass of small angular aperture may be made with its focus so near the anterior lens, that another glass of larger aperture could be constructed with a focus so far from the anterior lens that its *depth* of definition might exceed that of the objective of smaller aperture. Such a glass of large aperture would be, in every respect and for all purposes, greatly superior to the other

(*d*) But if an object-glass of moderate angular aperture, perfectly corrected, has its focus at the *greatest possible* distance from the anterior lens, every addition to the angular aperture, in another similar object-glass, necessarily (in the present state of science) requires the distance between the anterior lens and the object to be diminished Consequently, with the enlargement of the angular aperture, the *depth* of definition is generally diminished. If opticians could procure, for the manufacture of object-glasses, glass of such refractive and dispersive power as they would prefer, it would enable them to improve very much the depth of defi- nition in object-glasses of large angular aperture.

Some microscopists have had objectives of large angular aperture provided with a diaphragm, to be introduced behind the posterior lens, when viewing objects re- quiring greater extent of definition of parts above and below the focus. Some advantage may be gained in this manner, but object-glasses so arranged, cannot have as great depth of definition as if originally constructed with the greater distance between the anterior lens and the object, which would be possible with the limited aperture to which these glasses are thus temporarily reduced

In enlarging the angular aperture of our object-glasses, we have always sought to retain a reasonable working focus, or distance, between the anterior lens and the object, and so to combine depth of definition with the other qualities already enumerated, as shall give to our object-glasses the greatest degree of efficiency for the various uses of the practical microscopist.

PRICE LIST

OF

ACHROMATIC MICROSCOPES AND MICROSCOPICAL APPARATUS.

————— ◆◆◆ —————

PRICES OF MICROSCOPE STANDS,*

Exclusive of Objectives, Apparatus and Cases, except where mentioned.

No 1. **Educational Microscope,** (Sec. 45,) including
two eye-pieces, (Nos. 1 and 2,) and 1 inch and ¼ inch objectives of second class, which, by combination with the
two eye-pieces give magnifying powers respectively of 40,
70, 180, and 325 diameters, packed in a mahogany case, . **$45 00**

The same instrument, with the following additional apparatus,
(if ordered at the same time,) viz, Polariscope, Camera
lucida, Stage-micrometer, Bull's-eye condenser, Animalcule
cage, Stage-forceps, Hand-forceps, and a set of dissecting
instruments, ... **75 00**

No. 2. **Student's Microscope,** (Sec. 46,) two eye-pieces,
movable diaphragm, and 1 inch and ¼ inch objectives of
second class, (magnifying as above,) with mahogany case, . **50 00**

Three dissecting lenses, magnifying 5, 10 and 15 diameters respectively, mounted so as to be inserted in the arm of the
microscope when the body is removed, ... **8.00**

No. 3. **Student's Microscope,** (Sec. 47,) with the same
eye-pieces, objectives and diaphragm as the preceding, in a
mahogany case. This is the basis of a complete instrument, **60.00**

---

* All the Microscope Stands are so constructed that additional apparatus can be
supplied without requiring the instrument to be sent for fitting. It is only necessary to state the No of the instrument in the Catalogue, and the No engraved
upon it.

No 4, A. **Student's Larger Microscope Stand,** (Sec. 48,) with two eye-pieces, (Nos. 1 and 2,)   .   .   . $70 00

No. 4, B. The same stand and eye-pieces, with stage movable in two directions by rack and screw, instead of a lever, (Sec. 49,) 75.00

Either form of stage above mentioned, constructed so as to revolve around a steady centre, extra,   .   .   .   10.00

No. 4, C. The same stand and eye pieces as No. 4, A, with plain stage, ...   55.00

No. 5, A. **Student's Larger Microscope Stand,** made more portable than No. 4, and adapted to be used for dissecting; Lever stage revolving around a fixed centre, (See Sec. 50,) with two eye-pieces, (Nos. 1 and 2,)   .   .   .   90.00

No. 5, B. The same instrument as above, with stage movable in two directions by rack and screw, with two eye-pieces, .   95.00

No. 5, C. The same stand and eye-pieces, with plain stage, .   65.00

No. 6. **Portable Microscope Stand,** (Sec. 53,) with two eye-pieces carefully packed in a mahogany case of very convenient dimensions, with fittings for apparatus,   .   .   105.00

No 7, A. **Large Microscope Stand,** (Sec. 54,) with eye-pieces, Nos. 1 and 2, ...   125.00

No. 7, B. The same stand and eye-pieces as above, with stage movable in two directions by rack and screw, instead of a lever, ...   130.00

No. 8, A. **Inverted Microscope Stand,** (Sec. 56,) Condenser with rectangular prism, two eye-pieces and mahogany case, ...   60.00

No. 8, B. **Inverted Microscope Stand,** more complete than the former, combining in its construction all modern improvements.

Prof. Bailey's Indicator Stage, (Sec. 52) applied to No 4, C, extra, ...   40.00

The same applied to No 5, A, or to No. 7, A, instead of the lever stage with revolving motion, ...   15.00

Extra Compound Body with rectangular prism, Chevalier's plan, (Sec. 51,) to be substituted for the ordinary body in microscopes, Nos. 5 and 6, ...   18.00

Short elbow-tube, with rectangular prism, (Sec 51,) for the same purpose as the former, to fit into the draw-tube of Nos 4 and 7, ...   12 00

The same adapted to Nos. 1, 2 and 3, ...   10 00

Cases for the Microscopes. Upright mahogany case
for No. 7, with three drawers for object slides, &c, and
fitted for the reception of all accessories, ... $18 00
Upright mahogany case for Nos 4 and 5, with three drawers,
with fittings for all accessories,   .   .   .   15 00
Ditto of black walnut, and with one drawer,   .   .   .   12.00

Achromatic Object-Glasses for the Microscopes.

First Class. (See Section 58)

Object glasses	Angular Aperture about	Linear Magnifying Power (draw-tube closed) with Eye pieces.			Price
		No. 1	No. 2	No. 3	
2 inch.	13 degrees	20	35	60	$14 00
1 "	25	40	70	120	18.00
½ "	60	90	150	260	25 00
ditto without adjustment	50	"	"	"	20 00
¼ inch.	95–100	200	350	600	30 00
⅛ "	125–130	400	700	1100	40 00
1/12 "	160	590	1000	1600	60 00

Second Class. (See Section 61)

Object glasses	Angular Aperture about	Magnifying Powers as above	Price
2 inch.	10 degrees		$ 8 00
1 "	15		8.00
½ "	40		12 00
¼ "	65		15 00
⅛ "	90		25 00

Apparatus for Nos. 2 and 3.

SEE SECTION
    Movable Lever Stage, ...   10 00
84. Three Dark Wells and Holder,   .   .   .   2.50
110. Polariscope, (two Nicol's prisms,) .   .   .   14 00

Apparatus applied to Microscopes Nos. 4, 5 and 6.

SEE SECTION.

82.	Achromatic Condenser, ...	$27 00
82.	Brass work for ditto, alone, ...	7.50
79.	Movable Diaphragm Plate, ...	5.00
84.	Three Dark Wells and Holder, . . .	3.00
110.	Polariscope, (Nicol's prisms) . . .	20 00
118.	Ditto, with Revolving Selenite Carrier, . . .	23.00
69.	Prof. J. L. Smith's Eye-piece Micrometer and Goniometer, ...	15 00
83.	Nachet's Prism for oblique light, mounted with revolving motion, ...	12.00

Apparatus applied to Microscope No. 7.

82.	Achromatic Condenser, ...	25 00
82.	Brass work for ditto, alone, ...	8.00
110.	Polariscope, (Nicol's prisms,) . . .	20–25 00
118.	Ditto, with revolving selenite carrier, . .	25–30 00
84.	Three dark wells and holder, ...	3 50
69.	Prof. J. L. Smith's Eye-piece Micrometer and Goniometer, ...	20 00
83	Nachet's prism for oblique light, mounted with revolving motion, ...	12.00

Apparatus for the Inverted Microscope.

69.	Prof. J. L. Smith's Eye-piece Micrometer and Goniometer, ...	15.00
110.	Polariscope, (with two Nicol's prisms,) *specially adapted to the inverted microscope,* . . .	25 00
	Brass plate and spirit lamp for heating objects while under examination, ...	5 00

Apparatus for Microscopes in General.

63.	Glass Stage Micrometer, mounted in brass, $\frac{1}{100}$ to $\frac{1}{1000}$ of an inch, ...	4.00
64.	Cobweb-Micrometer Eye-piece, . . .	30 00
65.	Ross' Eye-piece Micrometer, ...	8.00
66.	Jackson's Micrometer with Eye-piece, . .	10 00
68.	Dr. White's Micrometer, ...	2.00

SEE SECTION.

73.	Fraunhofer's Stage Screw Micrometer, . . .	$40 00
74.	Wollaston's Camera lucida, ...	10.00
75	Nachet's Camera lucida, . . .	8.00
76.	Soemmering's Steel Speculum, . .	4 00
80.	Bull's eye condenser on stand, ...	6–7 00
81.	Smaller ditto, ...	5–6.50
84.	Lieberkuhn Speculum, ...	3 00
85.	Erector, ...	5 00
86	Orthoscopic Eye-piece, ...	10.00
87	Compressor, ...	6 00
88.	Animalcule cage with screw, ...	4 00
89.	Simple animalcule cage, ...	2 00
90.	Stage Forceps, ...	3 00
	Hand Forceps of brass, ...	75
	" " of steel, ...	50
91.	Frog Plate, ...	5 00
92.	Instrument for cutting circles of thin glass, . .	8 00
93.	Instrument for making cells of gold size, or other fluids, ...	3 00
	Extra Eye-piece, No. 1, 2 or 3, . . .	5 00
111.	Tourmalines fitted to Eye-piece, ...	4 00
117.	Darker's Selenite Stage, ...	10 00

WE ALSO MAKE THE FOLLOWING

Optical Instruments for various Scientific Researches.

Crown glass prism of any desired angle, mounted on a small brass stand, with two movements, ...	$8.00
Flint glass prism, mounted as above, ...	10–15.00
Prism of rock crystal, do.	15.00
Prism of Iceland spar, do.	15.00
Two prisms on one stand, to demonstrate the theory of Achromatism, ...	15 00
Hollow prism for fluids, with variable angle, on stand,	30 00
Biot's prism for volatile liquids, with perfectly plano-parallel sides of glass, ...	15 00
Ditto, with two compartments. . . .	20 00
Ditto, with three compartments, ...	25.00

Apparatus for viewing Fraunhofer's fixed dark lines in the solar spectrum, consisting of small achromatic telescope on stand, with flint glass prism and apparatus to mount the same before the objective, and a screen with fine slit, . $35 00

Apparatus for Fraunhofer's, Fresnel's and Schweid's experiments on diffraction, . . .   60 00

Wollaston's camera lucida, mounted for draftsmen, with clamp and the necessary adjustments, &c., from . . 10–18 00

Wollaston's goniometer, with telescope, graduated circle 5 inches diameter, the vernier gives the angles to single minutes, ... 40.00

Same, graduated circle 6 inches, with telescope and improved arrangement for holding and adjusting the crystals, . 60 00

Photometer, constructed by J & W. G., . . . 50.00

> This instrument was made first for Prof Silliman, Jr It is extremely sensitive, and the results of observations made with its aid are of a degree of accuracy and reliability which, we believe, has never before been obtained by a photometer

Dichroscopic lens, for discovering dichroism in crystals, . 6.00

Pocket achromatic simple microscope, for the use of botanists, mineralogists, and others, magnifying from 15 to 25 times linear, ... 6–8 00

Goniometer for measuring the angles of larger crystals, (Gambey's construction) This instrument may also be used for ascertaining the refractive power of prisms and crystals, ... 60 00

Norremberg's polariscope, . . .   30 00

Stauroscope, invented by Prof. V Kobell, of Munich, . 25.00

Biot's polariscope, mounted on tripod and column, and with graduated circles, ... 40.00

Soleil's apparatus for measuring the angles of the axes in biaxial crystals, and the diameter of the colored rings, . 60.00

Soleil's sacharimeter, ... 90.00

Apparatus for showing the transient polarizing structure of glass plates by unequal heating, . . .   2.00

Apparatus for producing the polarizing structure by bending,   4.00

Apparatus for producing the same by compression, . . 3.00

Nicol's prisms, according to the size, from . . . 4–10.00

WE ALSO KEEP FOR SALE

Raspail's Simple Dissecting Microscopes with three lenses, . . $12 00

Glass slips for mounting Microscopic preparations, per dozen,   36 00

Square covers of thin glass, per dozen, . 25

Round covers of thin glass, per dozen, . 36

Microscopic specimens at various prices

CPSIA information can be obtained
at www.ICGtesting.com
Printed in the USA
LVHW080827190323
741893LV00056B/1387